basic guide

to

Helicopters

DRAKE PUBLISHERS INC. NEW YORK•LONDON

Published in 1978 by
Drake Publishers, Inc.
801 Second Avenue
New York, N.Y. 10017

Library of Congress Cataloging in Publication Data

United States. Flight Standards Service.
    Basic guide to helicopters.

    Reprint of the rev. ed. published in 1973
under title: Basic helicopter handbook.
    Bibliography: p.
    Includes index.
    1. Helicopters—Piloting—Handbooks, manuals,
etc. I. Title.
TL716.5.U53   1978   629.133'35   78-7077
ISBN 0-8473-1756-0

CONTENTS

ILLUSTRATIONS

Some illustrations may show exaggerated positions of the helicopter or rotor system to better represent a situation.

FOREWORD

With the advent of light and relatively inexpensive helicopters, interest in helicopter flying has increased tremendously as evidenced by the steady increase in the number of applicants for pilot certificates with a helicopter rating. This increase in interest has given added emphasis to the need for adequate helicopter training materials available to the general aviation public.

Generally, current helicopter training manuals consist of two types — those published by the military services and those published by various helicopter manufacturers for their own particular helicopters. Both are excellent publications but the former are not available to the public and the latter, for the most part, are applicable only to a particular helicopter.

This handbook, developed by Flight Standards Service of the Federal Aviation Administration, was prepared as a technical manual for applicants preparing for private, commercial, and flight instructor pilot certificates with a helicopter rating. Currently certificated helicopter flight instructors will also find it valuable as an aid in training students since it gives detailed coverage of helicopter aerodynamics, performance, and flight maneuvers. It does not include material on such things as weather, navigation, radio navigation and communications, use of a computer, use of flight information publications, and pertinent Federal Aviation Regulations. These common topics, with the exception of regulations, may be found in the FAA publication, *Pilot's Handbook of Aeronautical Knowledge,* and in other FAA publications listed in the references. Up-to-date applicable regulations should be purchased and studied by the applicant.

Chapter 1. GENERAL AERODYNAMICS

Unless otherwise indicated, this handbook is based on a helicopter with the following characteristics:

1—An unsupercharged reciprocating engine.

2—A single main rotor rotating in a counterclockwise direction (looking downward on the rotor).

3—An antitorque (tail) rotor.

4—Skid-type landing gear.

Information is intended to be general in nature and should apply to most helicopters having these characteristics.

Before launching into a detailed discussion of the various forces acting on a helicopter in flight, it is first necessary that you understand the meaning of a few basic aerodynamic terms, how the force of lift is created, and the effect that certain factors have on lift.

Airfoil.—An airfoil is any surface designed to produce lift or thrust when air passes over it. The wings of airplanes are airfoils, also the propellers. Airfoils on a helicopter are the rotor blades. The wing of an airplane is normally an unsymmetrical airfoil, that is, the top surface has more curvature than the lower surface.

The main rotor blades of most helicopters are symmetrical airfoils; that is, having the same curvature on both upper and lower surfaces (fig. 1). Much research, however, is being conducted in the use of unsymmetrical airfoils for main rotor blades, and at least one currently manufactured make of helicopter is equipped with main rotor blades that are not considered true symmetrical airfoils.

On an unsymmetrical airfoil, the center of pressure is variable—as the angle of attack increases, the center of

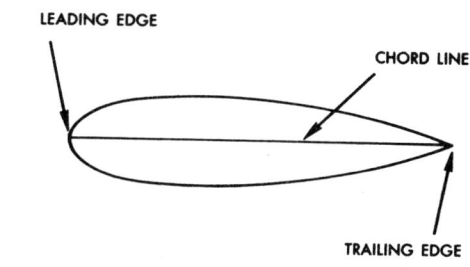

FIGURE 2.—Chord line of an airfoil is the imaginary line joining the leading and trailing edges of the airfoil.

pressure moves forward along the airfoil surface; as the angle of attack decreases, the center of pressure moves rearward. On a symmetrical airfoil, center of pressure movement is very limited. A symmetrical airfoil is preferred for rotor blades so that a relatively stable center of pressure is maintained. Improvements in control systems may allow more latitude in blade designs in the future.

Chord line.—The chord line of an airfoil is an imaginary straight line from the leading edge to the trailing edge of the airfoil (fig. 2).

Relative wind.—Relative wind is the direction of the airflow with respect to an airfoil. If an airfoil moves forward horizontally, the relative wind moves backward horizontally (fig. 3). If an airfoil moves backward horizontally, the relative wind moves forward horizontally. If an airfoil moves forward and upward, the relative wind moves backward and downward. If an airfoil moves backward and downward, the relative wind moves forward and upward. Thus, the flight path and relative wind are parallel but travel in opposite directions. (Forward and backward as used here are relative to the fore and aft axis of the helicopter—forward meaning in the direction that the nose of the helicopter points, and backward meaning the direction the tail points.)

Relative wind may be affected by several factors including the rotation of the rotor blades, horizontal movement

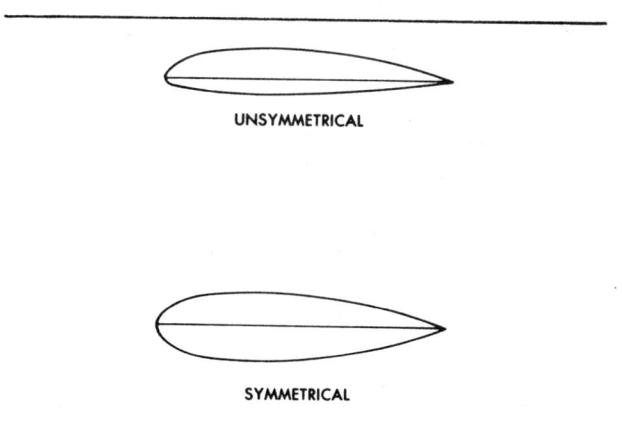

FIGURE 1.—Symmetrical and unsymmetrical airfoils.

1

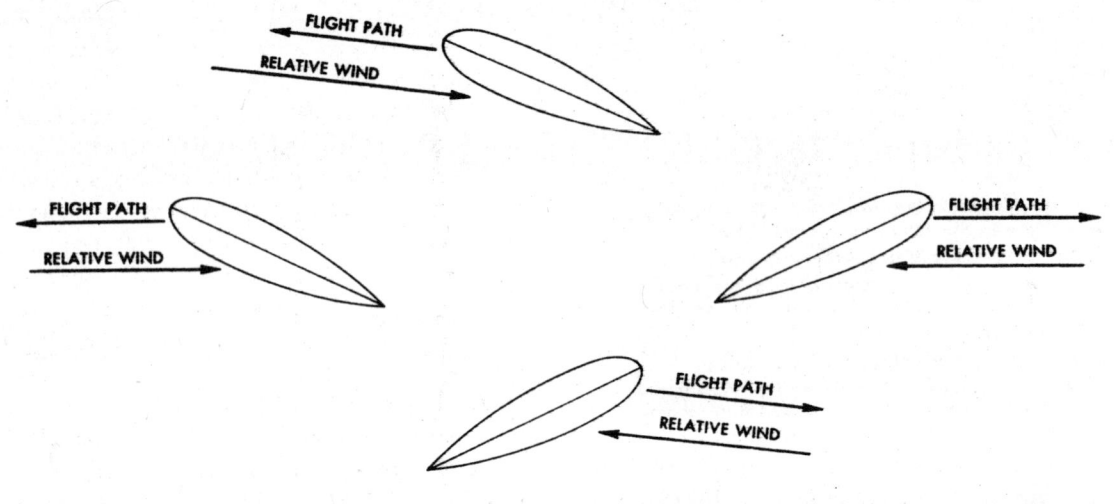

FIGURE 3.—Relationship between the flight path of an airfoil and relative wind. Relative wind is parallel and in the opposite direction to the flight path.

of the helicopter, flapping of the rotor blades, and wind speed and direction.

Relative wind is created by the motion of an airfoil through the air, by the motion of air past an airfoil, or by a combination of the two. For a helicopter, the relative wind is the flow of air with respect to the rotor blades. When the rotor is stopped, wind blowing over the blades creates a relative wind; when the helicopter is hovering in a no-wind condition, relative wind is created by the motion of the rotor blades through the air; when the helicopter is hovering in a wind, the relative wind is a combination of the wind and the motion of the rotor blades through the air; and when the helicopter is in horizontal flight, the relative wind is a combination of the rotation of the rotor blades and the movement of the helicopter.

Pitch angle.—The rotor blade pitch angle is the acute angle between the blade chord line and a reference plane determined by the main rotor hub. Since the rotor plane of rotation is parallel to the plane containing the main rotor hub, the rotor blade pitch angle could also be described as the acute angle between the blade chord line and the rotor plane of rotation (fig. 4). The pitch angle can be varied by the pilot through the use of cockpit controls (collective and cyclic pitch controls) provided for this purpose.

Angle of attack.—The angle of attack is the angle between the chord line of the airfoil and the direction of the relative wind (fig. 5). The angle of attack should not be confused with the pitch angle of the rotor blades. The pitch angle is determined by the position of the appro-

priate cockpit controls, (collective and cyclic pitch), whereas the angle of attack is determined by the direction of the relative wind. The angle of attack may be less than, equal to, or greater than the pitch angle as shown in figure 6. The pilot can increase or decrease the angle of attack by changing the pitch angle of the rotor blades. If the pitch angle is increased, the angle of attack is increased; if the pitch angle is decreased, the angle of attack is decreased. Since the angle of attack is dependent on the relative wind, the same factors that affect the relative wind also affect the angle of attack.

Lift.—The force, lift, is derived from an airfoil through a principle often referred to as Bernoulli's Principle or the

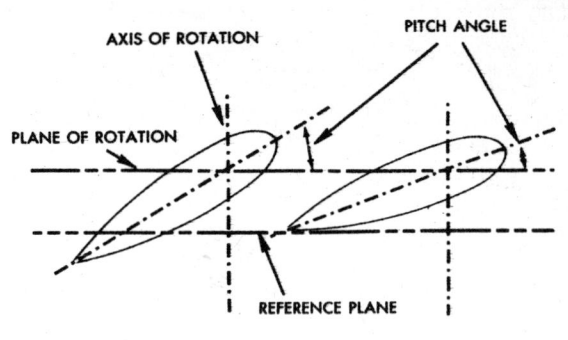

FIGURE 4.—The pitch angle of a rotor blade is the angle between the chord line and a reference plane determined by the rotor hub or the plane of rotation.

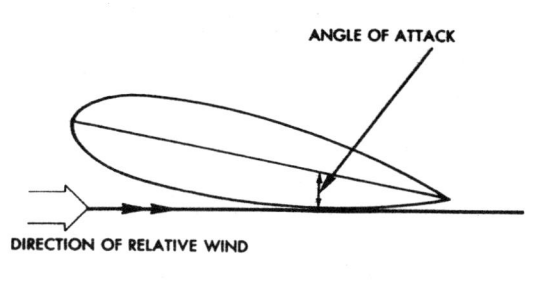

ANGLE OF ATTACK

DIRECTION OF RELATIVE WIND

FIGURE 5.—Angle of attack is the angle between the relative wind line and the chord line.

"venturi effect." As air velocity increases through the constricted portion of a venturi tube, the pressure decreases. Compare the upper surface of an airfoil with the constriction in the venturi tube (fig. 7). They are very similar. The upper half of the venturi tube is replaced by layers of undisturbed air. Thus, as air flows over the upper surface of an airfoil, the curvature of the airfoil causes an increase in the speed of the airflow. The increased speed of airflow results in a decrease in pressure on the upper surface of the airfoil. At the same time, airflow strikes the lower surface of the airfoil at an angle, building up pressure. The combination of decreased pressure on the upper surface and increased pressure on the lower surface results in an upward force. This is the force, lift.

Drag (airfoil).—At the same time the airfoil is producing lift, it also is subject to a drag force. Drag is the term used for the force that tends to resist movement of the airfoil through the air—the retarding force of inertia and wind resistance. It acts parallel and in the opposite direction to the movement of the airfoil or, if you prefer, in the same direction as the relative wind. It is this force that causes a reduction in rotor RPM (revolutions per minute) when the angle of attack is increased. An increase in angle of attack then not only produces an increase in lift, but it also produces an increase in drag (fig. 8).

Stall.—When the angle of attack increases up to a certain point, the air can no longer flow smoothly over the top surface because of the excessive change of direction required. This loss of streamlined flow results in a swirling, turbulent airflow and a large increase in drag. The turbulent airflow also causes a sudden increase in pressure on the top surface resulting in a large loss of lift. At this point, the airfoil is said to be in a stalled condition.

Lift and angle of attack.—As the angle of attack of an airfoil increases, the lift increases (up to the stall angle) providing the velocity of the airflow (relative wind) remains the same (fig. 8). Since the pilot can increase or decrease the angle of attack by increasing or decreasing the pitch angle of the rotor blades through the use of the collective pitch cockpit control, then he can increase or decrease the lift produced by the rotor blades. He must remember, however, that any increase in angle of attack

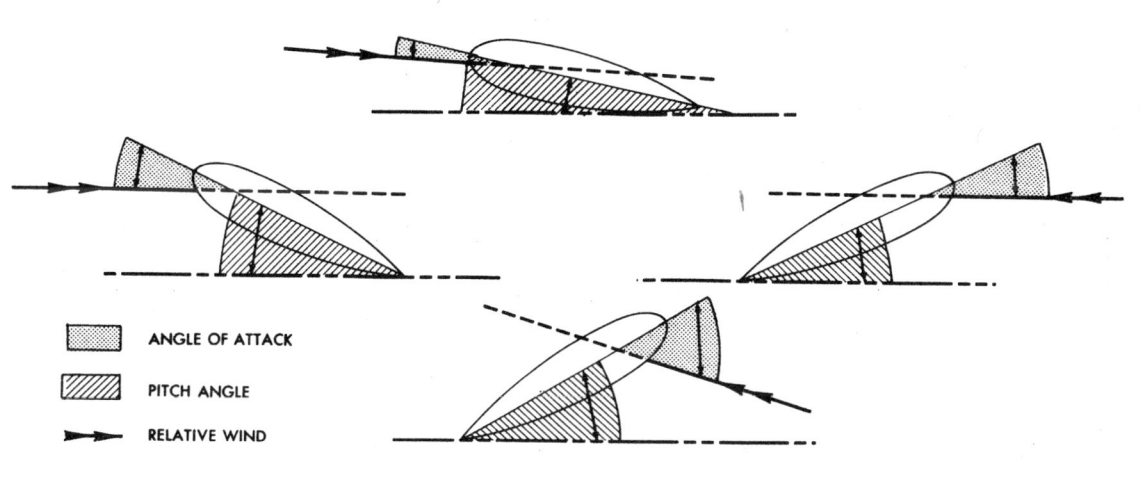

ANGLE OF ATTACK

PITCH ANGLE

RELATIVE WIND

FIGURE 6.—The relationship between the angle of attack and pitch angle for various positions of the rotor blade in the plane of rotation during forward flight. Angle of attack is less than pitch angle 90° to pilot's right (top); greater than pitch angle 90° to pilot's left (bottom); and equal to pitch angle in the fore (left) and aft (right) positions.

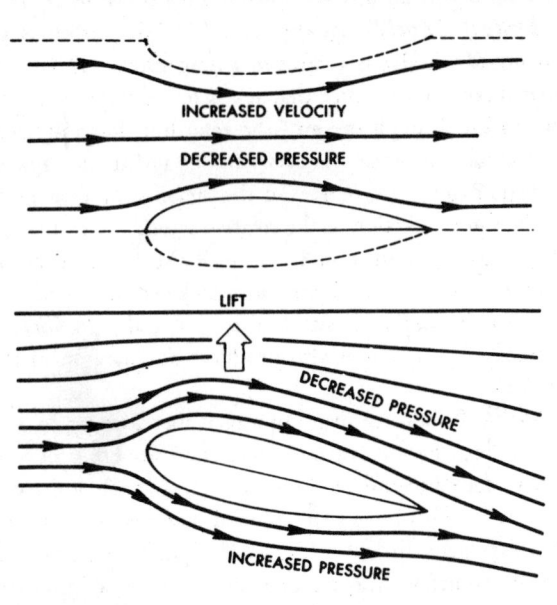

INCREASED VELOCITY

DECREASED PRESSURE

LIFT

DECREASED PRESSURE

INCREASED PRESSURE

FIGURE 7.—(Top) Bernoulli's Principle: Increased air velocity produces decreased pressure; (Bottom) Lift is produced by an airfoil through a combination of decreased pressure above the airfoil (as per Bernoulli's Principle), and increased pressure beneath.

will also increase drag on the rotor blades tending to slow down the rotor rotation. Additional power will be required to prevent this slowing down of the rotor.

Lift and velocity of airflow.—As the velocity of the airflow (relative wind) increases, the lift increases for any given angle of attack. Since the pilot can increase or decrease the rotor RPM which, in turn, increases or decreases the velocity of the airflow, he can change the amount of lift. As a general rule, however, the pilot attempts to maintain a constant rotor RPM and changes the lift force by varying the angle of attack.

Lift and air density.—Lift varies directly with the density of the air—as the air density increases, lift and drag increase; as air density decreases, lift and drag decrease.

What affects air density? Altitude and atmospheric changes affect air density. *The higher the altitude the less dense is the air.* At 10,000 feet the air is only two-thirds as dense as the air at sea level. Therefore, if a helicopter is to maintain its lift, the angle of attack of the rotor blades must be increased. In order to increase the angle of attack, the pilot must increase the pitch angle of the blades. We have already seen that, as the pitch angle increases, drag on the rotor system increases and the rotor RPM tends to decrease. Therefore,

more power must be applied to prevent a decrease in rotor RPM. This is why a helicopter requires more power to hover at higher altitudes than under the same conditions at lower altitudes. (See fig. 52 and the accompanying discussion.)

Due to the atmospheric changes in temperature, pressure, or humidity, the density of the air may be different, even at the same altitude, from one day to the next or from one location in the country to another. Because air expands when heated, *hot air is less dense than cold air.* In order for the helicopter to produce the same amount of lift on a hot day as on a cold day, the rotor blades must be operated at a higher angle of attack. This requires that the blades be operated at a greater pitch angle which increases the drag and tends to reduce rotor RPM. Therefore to maintain a constant rotor RPM, more throttle is required. For this reason, a helicopter requires more power to hover on a hot day than on a cold day. (See fig. 53 and the accompanying discussion.)

Because air expands as pressure is decreased, there will be fluctuations in the air density due to changes in atmospheric pressure. *The lower the pressure, the less dense the air* and, for the same reason stated previously, the greater the power required to hover.

Because water vapor weighs less than an equal amount of dry air, *moist air (high relative humidity) is less dense than dry air (low relative humidity).* Because of this, a

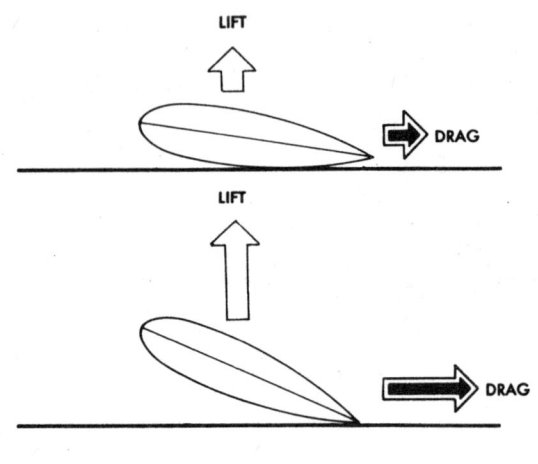

LIFT

DRAG

LIFT

DRAG

FIGURE 8.—Relationship between angle of attack and lift and drag forces. As the angle of attack increases, lift and drag increase.

helicopter will require more power to hover on a humid day than on a dry day. (See fig. 54 and the accompanying discussion.) This is especially true on hot, humid days because *the hotter the day, the greater the amount of water vapor the air can hold*. The more moisture (water vapor) in the air, the less dense the air.

From the above discussion, it is obvious that *a pilot should beware of high, hot, and humid conditions*—high altitudes, hot temperatures, and high moisture content. (See fig. 55 and the accompanying discussion.) He should be especially aware of these conditions at his destination, since sufficient power may not be available to complete a landing safely, particularly when the helicopter is operating at high gross weights. (See fig. 64 and the accompanying discussion.)

Lift and weight.—The total weight (gross weight) of a helicopter is the first force that must be overcome before flight is possible. Lift, the force which overcomes or balances the force of weight, is obtained from the rotation of the main rotor blades.

Thrust and drag.—Thrust moves the aircraft in the desired direction; drag, the retarding force of inertia and wind resistance, tends to hold it back. In vertical flight, drag acts downward; in horizontal flight, drag acts horizontally and opposite in direction to the thrust component. Thrust, like lift, is obtained from the main rotor. Drag, as discussed here, is the drag of the entire helicopter—not just the drag of the rotor blades which was discussed earlier. The use of the term "drag" in subsequent portions of this handbook should be considered as having this same connotation. In future references to the drag of the rotor blades, the statement "drag of the rotor blades, or rotor system" will be used.

Chapter 2. AERODYNAMICS OF FLIGHT

POWERED FLIGHT

In any kind of flight (hovering, vertical, forward, side-ward, or rearward), the total lift and thrust forces of a rotor are perpendicular to the tip-path plane or plane of rotation of the rotor (fig. 9). The tip-path plane is the imaginary circular plane outlined by the rotor blade tips in making a cycle of rotation.

Forces acting on the helicopter

During any kind of horizontal or vertical flight, there are four forces acting on the helicopter—lift, thrust, weight, and drag. Lift is the force required to support the weight of the helicopter. Thrust is the force required to overcome the drag on the fuselage and other helicopter components.

Hovering flight.—During hovering flight in a no-wind condition, the tip-path plane is horizontal, that is, parallel to the ground. Lift and thrust act straight up; weight and drag act straight down. The sum of the lift and thrust forces must equal the sum of the weight and drag forces in order for the helicopter to hover.

Vertical flight.—During vertical flight in a no-wind condition, the lift and thrust forces both act vertically upward. Weight and drag both act vertically downward. When lift and thrust equal weight and drag, the helicopter hovers; if lift and thrust are less than weight and drag, the helicopter descends vertically; if lift and thrust are greater than weight and drag, the helicopter rises vertically (fig. 10).

Forward flight.—For forward flight, the tip-path plane is tilted forward, thus tilting the total lift–thrust force forward from the vertical. This resultant lift–thrust force can be resolved into two components—lift acting vertically upward and thrust acting horizontally in the direction of flight. In addition to lift and thrust, there are weight, the downward acting force, and drag, the rearward acting or retarding force of inertia and wind resistance (fig. 11).

In straight-and-level, unaccelerated forward flight, lift equals weight and thrust equals the drag. (Straight-and-level flight is flight with a constant heading and at a constant altitude.) If lift exceeds weight, the helicopter climbs; if the lift is less than weight, the helicopter descends. If thrust exceeds drag, the helicopter speeds up; if thrust is less than drag, it slows down.

Sideward flight.—In sideward flight, the tip-path plane is tilted sideward in the direction that flight is desired

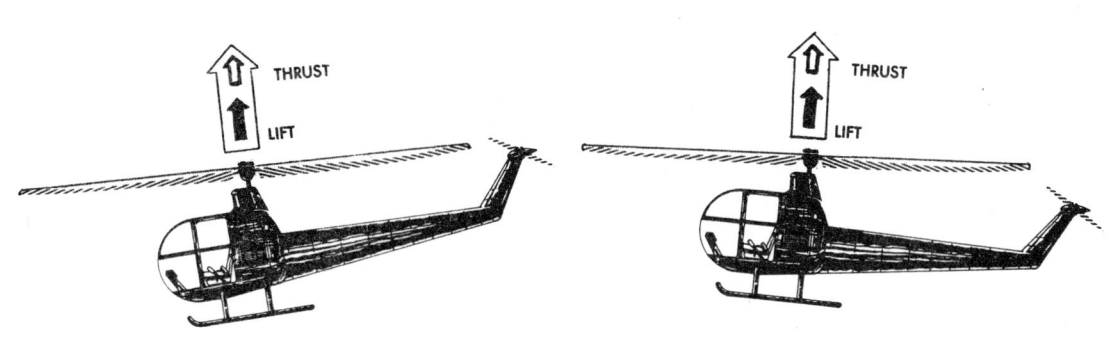

FIGURE 9.—The total lift–thrust force acts perpendicular to the rotor disc or tip-path plane.

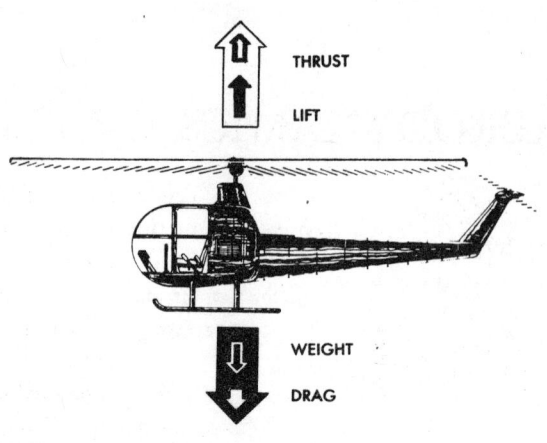

THRUST

LIFT

WEIGHT

DRAG

VERTICAL ASCENT

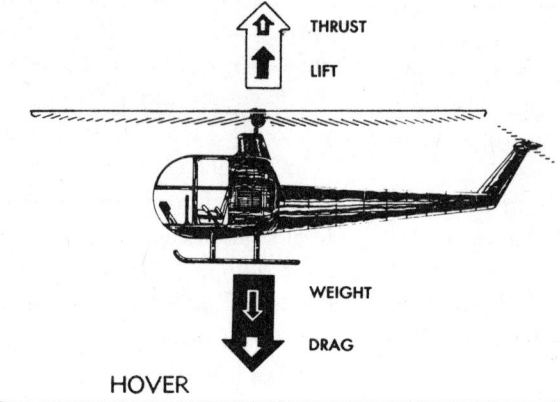

THRUST

LIFT

WEIGHT

DRAG

HOVER

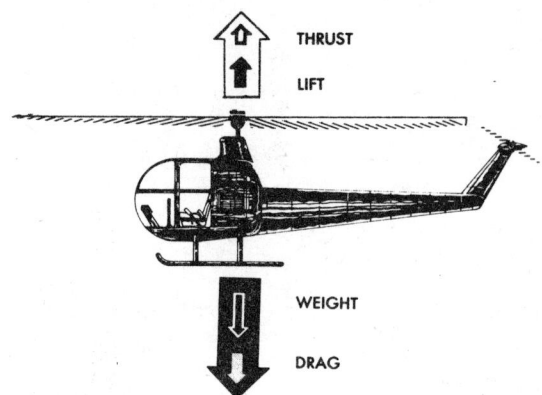

THRUST

LIFT

WEIGHT

DRAG

VERTICAL DESCENT

FIGURE 10.—Forces acting on the helicopter during a hover and vertical flight.

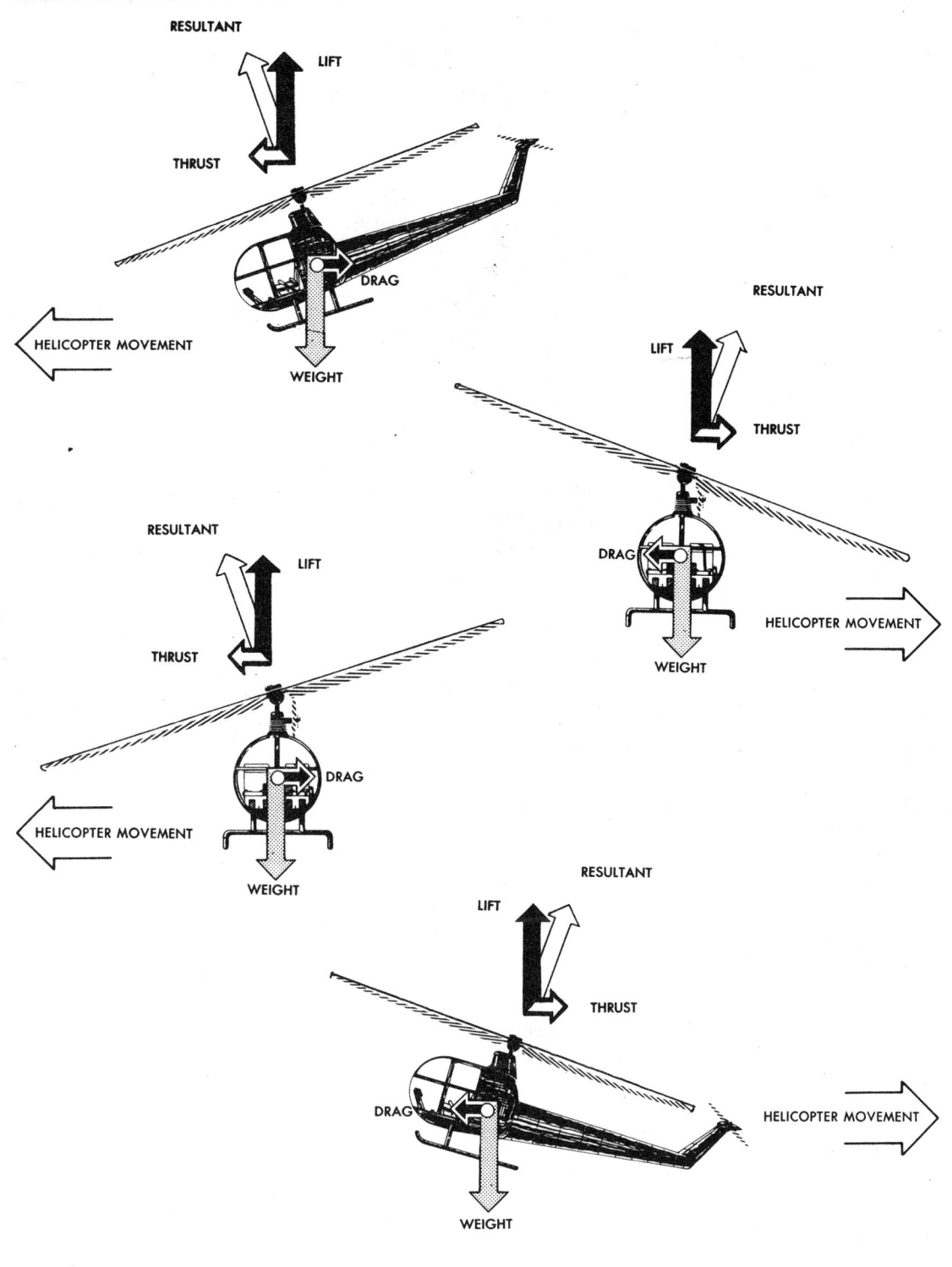

FIGURE 11.—Forces acting on the helicopter during forward, sideward, and rearward flight.

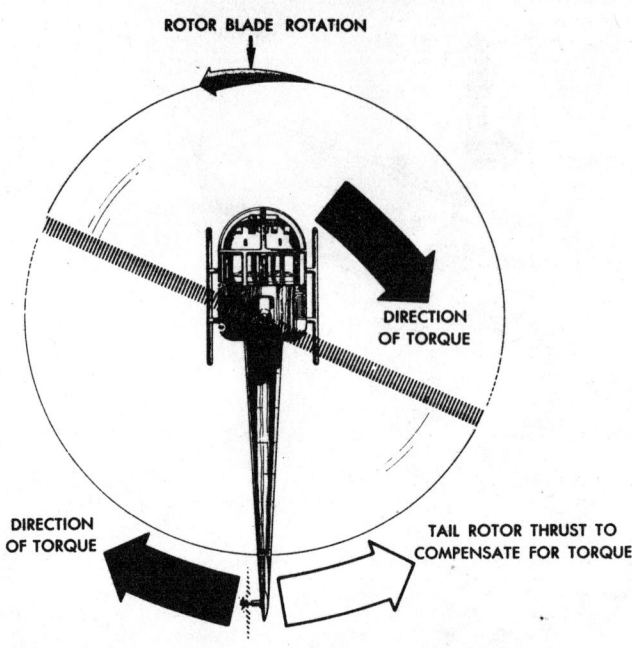

ROTOR BLADE ROTATION

DIRECTION
OF TORQUE

DIRECTION
OF TORQUE

TAIL ROTOR THRUST TO
COMPENSATE FOR TORQUE

FIGURE 12.—Tail rotor thrust compensates for the torque effect of the main rotor.

thus tilting the total lift–thrust vector sideward. In this case, the vertical or lift component is still straight up, weight straight down, but the horizontal or thrust component now acts sideward with drag acting to the opposite side (fig. 11).

Rearward flight.—For rearward flight, the tip-path plane is tilted rearward tilting the lift–thrust vector rearward. The thrust component is rearward and drag forward, just the opposite to forward flight. The lift component is straight up and weight straight down (fig. 11).

Torque.—Newton's third law of motion states, "To every action there is an equal and opposite reaction." As the main rotor of a helicopter turns in one direction, the fuselage tends to rotate in the opposite direction (fig. 12). This tendency for the fuselage to rotate is called torque. Since torque effect on the fuselage is a direct result of engine power supplied to the main rotor, any change in engine power brings about a corresponding change in torque effect. The greater the engine power, the greater the torque effect. Since there is no engine power being supplied to the main rotor during autorotation, there is no torque reaction during autorotation.

Auxiliary rotor.—The force that compensates for torque and keeps the fuselage from turning in the direction opposite to the main rotor is produced by means of an auxiliary rotor located on the end of the tail boom. This auxiliary rotor, generally referred to as a *tail rotor,*

or antitorque rotor, produces thrust in the direction opposite to torque reaction developed by the main rotor (fig. 12). Foot pedals in the cockpit permit the pilot to increase or decrease tail-rotor thrust, as needed, to neutralize torque effect.

Gyroscopic precession.—The spinning main rotor of a helicopter acts like a gyroscope. As such, it has the properties of gyroscopic action, one of which is precession. Gyroscopic precession is the resultant action or deflection of a spinning object when a force is applied to this object. This action occurs approximately 90° in the direction of rotation from the point where the force is applied (fig. 13). Through the use of this principle, the tip-path plane of the main rotor may be tilted from the horizontal.

The movement of the cyclic pitch control in a two-bladed rotor system increases the angle of attack of one rotor blade with the result that a greater lifting force is applied at this point in the plane of rotation. This same control movement simultaneously decreases the angle of attack of the other blade a like amount thus decreasing the lifting force applied at this point in the plane of rotation. The blade with the increased angle of attack tends to rise; the blade with the decreased angle of attack tends to lower. However, because of the gyroscopic precession property, the blades do not rise or lower to maximum deflection until a point approximately 90° later in the

10

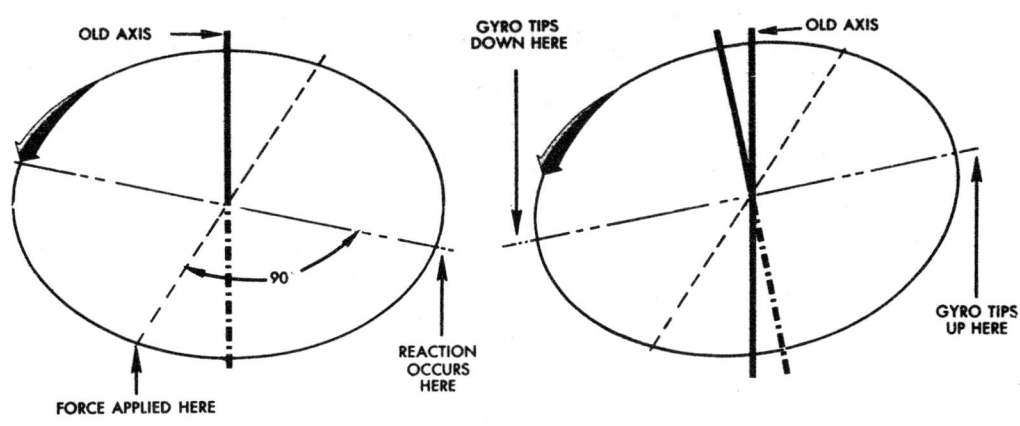

FIGURE 13.—Gyroscopic Precession Principle: When a force is applied to a spinning gyro, the maximum reaction occurs 90° later in the direction of rotation.

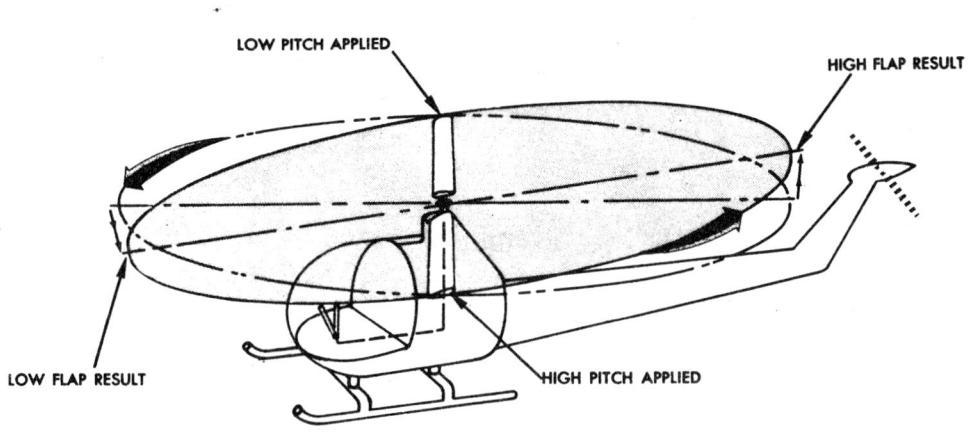

FIGURE 14.—Rotor disc acts like a gyro. When a rotor blade pitch change is made, maximum reaction occurs approximately 90° later in the direction of rotation.

plane of rotation. In the illustration (fig. 14), the retreating blade angle of attack is increased and the advancing blade angle of attack is decreased resulting in a tipping forward of the tip-path plane, since maximum deflection takes place 90° later when the blades are at the rear and front respectively.

In a three-bladed rotor, the movement of the cyclic pitch control changes the angle of attack of each blade an appropriate amount so that the end result is the same—a tipping forward of the tip-path plane when the maximum change in angle of attack is made as each blade passes the same points at which the maximum increase and decrease are made in the illustration (fig. 14) for the two-bladed rotor. As each blade passes the 90° position on the left, the maximum increase in angle of attack occurs. As each blade passes the 90° position to the right, the maximum decrease in angle of attack occurs. Maximum deflection takes place 90° later—maximum upward deflection at the rear and maximum downward deflection at the front—and the tip-path plane tips forward.

Dissymmetry of lift.—The area within the tip-path plane of the main rotor is known as the *disc area* or *rotor disc*. When hovering in still air, lift created by the rotor blades at all corresponding positions around the rotor disc

11

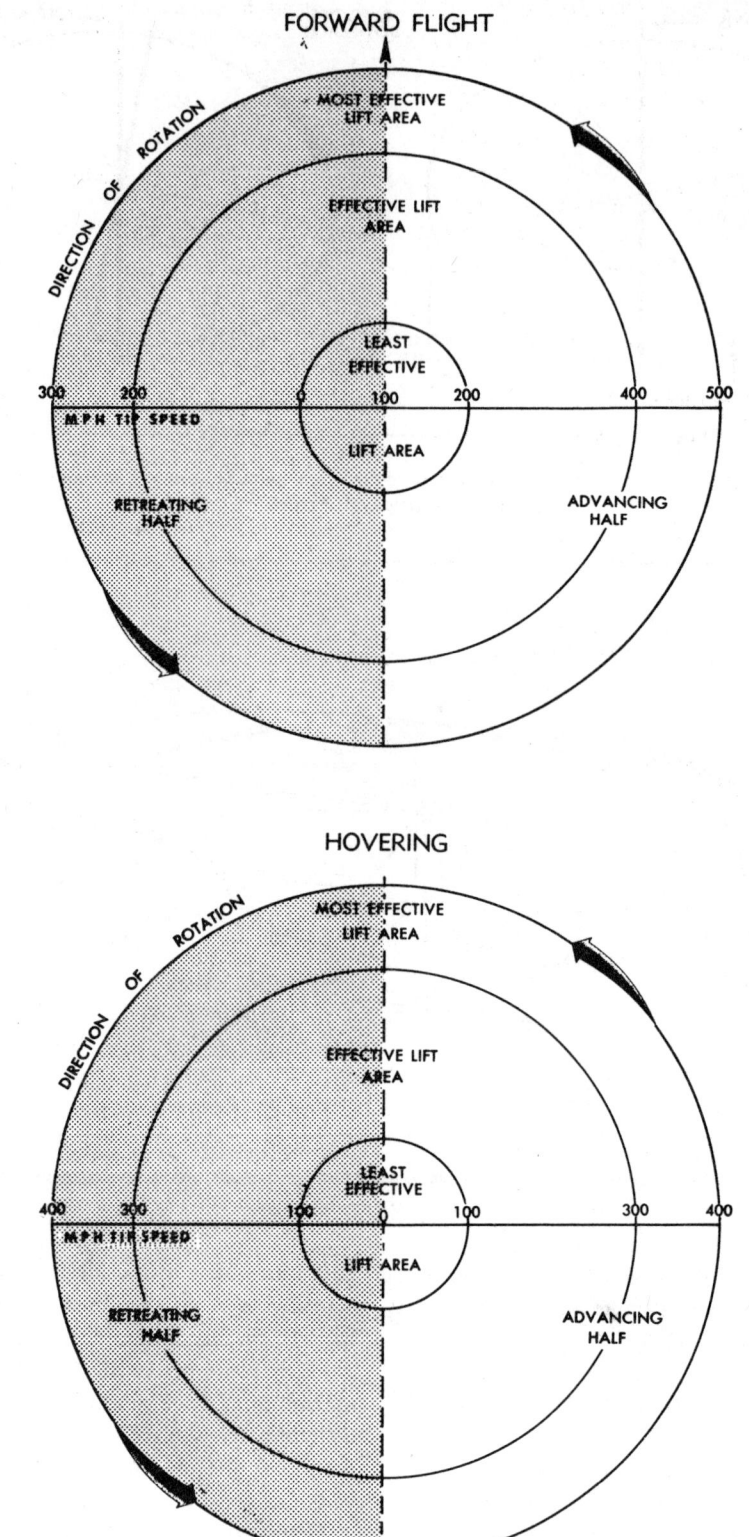

FIGURE 15.—Comparison of rotor blade speeds for the advancing blade and retreating blade during hovering and forward flight.

is equal. Dissymmetry of lift is created by horizontal flight or by wind during hovering flight, and is the difference in lift that exists between the advancing blade half of the disc area and the retreating blade half.

At normal rotor operating RPM and zero airspeed, the rotating blade-tip speed of most helicopter main rotors is approximately 400 miles per hour. When hovering in a no-wind condition, the speed of the relative wind at the blade tips is the same throughout the tip-path plane (fig. 15 bottom). The speed of the relative wind at any specific point along the rotor blade will be the same throughout the tip-path plane; however, the speed is reduced as this point moves closer to the rotor hub as indicated by the two inner circles. As the helicopter moves into forward flight, the relative wind moving over each rotor blade becomes a combination of the rotational speed of the rotor and the forward movement of the helicopter (fig. 15 top). At the 90° position on the right side, the advancing blade has the combined speed of the blade velocity plus the speed of the helicopter. At the 90° position on the left side, the retreating blade speed is the blade velocity less the speed of the helicopter. (In the illustration, the helicopter is assumed to have a forward airspeed of 100 miles per hour.) In other words, the relative wind speed is at a maximum at the 90° position on the right side and at a minimum at the 90° position on the left side.

Earlier in this handbook, the statement was made that for any given angle of attack, lift increases as the velocity of the airflow over the airfoil increases. It is apparent that the lift over the advancing blade half of the rotor disc will be greater than the lift over the retreating blade half during horizontal flight or when hovering in a wind unless some compensation is made. It is equally apparent that the helicopter will roll to the left unless some compensation is made. What compensation is made to equalize the lift over the two halves of the rotor disc?

Blade flapping.—In a three-bladed rotor system, the rotor blades are attached to the rotor hub by a horizontal hinge which permits the blades to move in a vertical plane, i.e., flap up or down, as they rotate (fig. 16). In forward flight and assuming that the blade-pitch angle remains constant, the increased lift on the advancing blade will cause the blade to flap up decreasing the angle of attack because the relative wind will change from a horizontal direction to more of a downward direction. The decreased lift on the retreating blade will cause the blade to flap down increasing the angle of attack because the relative wind changes from a horizontal direction to more of an upward direction (fig. 3). The combination of decreased angle of attack on the advancing blade and increased angle of attack on the retreating blade through blade flapping action tends to equalize the lift over the two halves of the rotor disc.

In a two-bladed system, the blades flap as a unit. As the advancing blade flaps up due to the increased lift, the retreating blade flaps down due to the decreased lift. The change in angle of attack on each blade brought about by this flapping action tends to equalize the lift over the two halves of the rotor disc.

The position of the cyclic pitch control in forward flight also causes a decrease in angle of attack on the advancing blade and an increase in angle of attack on the retreating blade. This together with blade flapping equalizes lift over the two halves of the rotor disc.

Coning.—Coning is the upward bending of the blades caused by the combined forces of lift and centrifugal force. Before takeoff, the blades rotate in a plane nearly perpendicular to the rotor mast, since centrifugal force is the major force acting on them (fig. 17).

As a vertical takeoff is made, two major forces are acting at the same time—centrifugal force acting outward perpendicular to the rotor mast and lift acting upward and parallel to the mast. The result of these two forces is that the blades assume a conical path instead of remaining in the plane perpendicular to the mast (fig. 17).

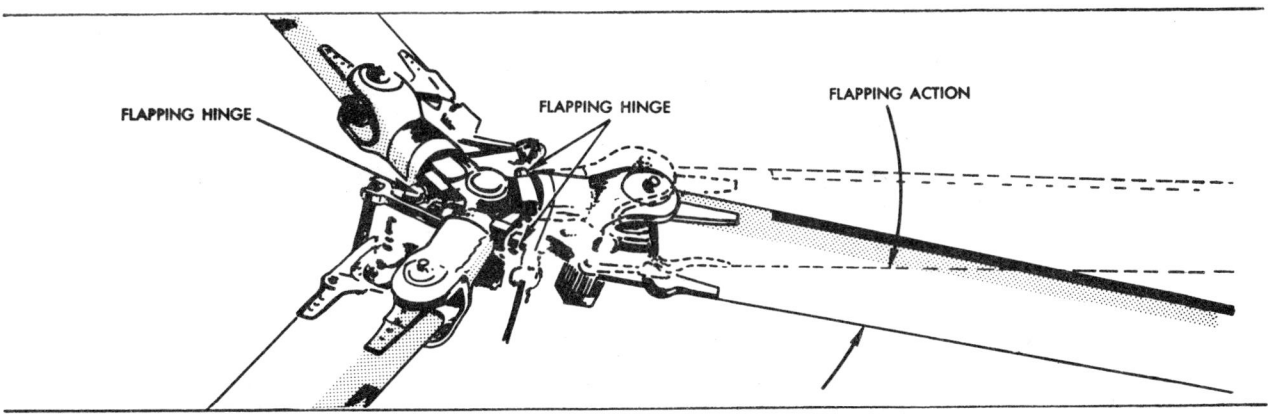

FIGURE 16.—Flapping action about the flapping hinges. Drag hinges can also be seen.

Coning results in blade bending in a semirigid rotor; in an articulated rotor, the blades assume an upward angle through movement about the flapping hinges.

Axis of rotation.—The axis of rotation of a helicopter rotor is the imaginary line about which the rotor rotates. It is represented by a line drawn through the center of, and perpendicular to, the tip-path plane. It is not to be confused with the rotor mast. The only time the rotor axis of rotation coincides with the rotor mast is when the tip-path plane is perpendicular to the rotor mast (fig. 18).

Coriolis effect.—In a three-bladed rotor system, when a rotor blade flaps upward, the distance of the center of mass of the blade from the axis of rotation decreases (fig. 19). (Keep in mind that, due to coning, the rotor blades will not flap below a plane passing through the rotor hub and perpendicular to the axis of rotation.) The distance of the center of mass from the axis of rotation (measured perpendicular to the axis of rotation) times the rotational velocity must always remain the same for a given rotor RPM. Since this distance becomes shorter when the blade flaps upward (fig. 19), the rotational velocity must increase for the product of the two to remain the same; conversely, when the blade flaps downward, the blade rotation must slow down, since the center of mass is moved farther from the axis of rotation. This change in blade velocity in the plane of rotation causes a hunting action about the vertical (drag) hinge. This tendency of the blade to increase or decrease its velocity is known as coriolis effect. The acceleration or decelera-

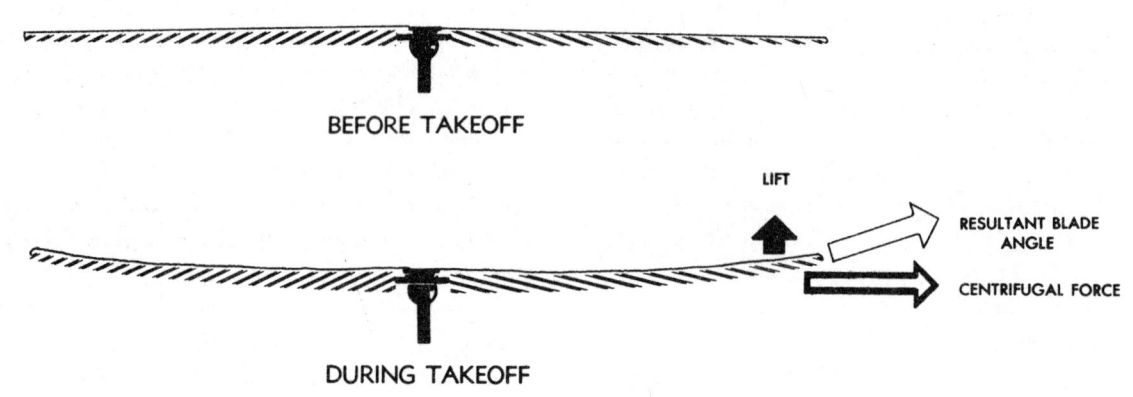

FIGURE 17.—Blade coning is a result of lift and centrifugal force.

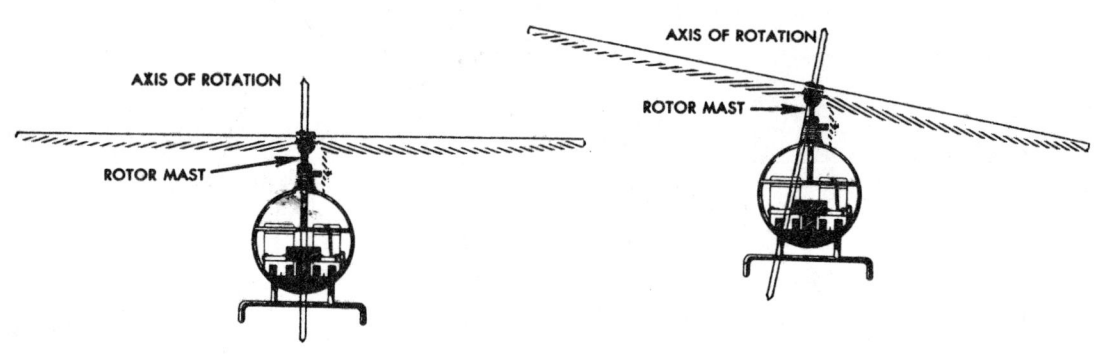

FIGURE 18.—The axis of rotation is the imaginary line about which the rotor rotates and is perpendicular to the tip-path plane.

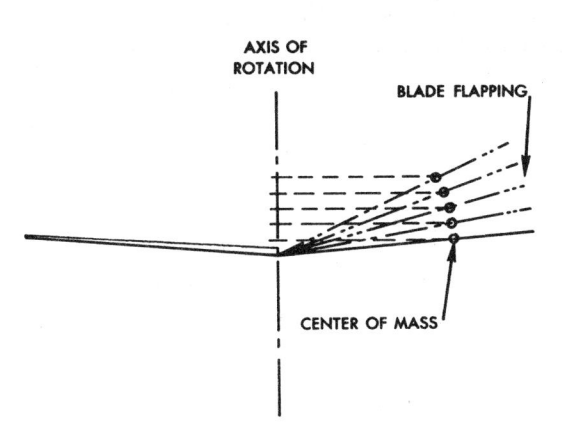

FIGURE 19.—Coriolis effect is the change in blade velocity to compensate for the change in distance of the center of mass from the axis of rotation as the blades flap.

tion is absorbed by dampers or the blade structure itself depending on the design.

Coriolis effect might be compared to a spinning skater. When the skater extends her arms, her rotation slows down because the center of mass moves farther from the axis of rotation. When she pulls in her arms, the rotation speeds up because the center of mass moves closer to the axis of rotation.

Two-bladed rotor systems are normally subject to coriolis effect to a much lesser degree since the blades are generally "underslung" with respect to the rotor hub and the change in the distance of the center of mass from the axis of rotation is small. The hunting action is absorbed by the blades through bending. If a two-bladed rotor system is not "underslung" then it will be subject to coriolis effect comparable to a fully articulated system.

Translating tendency or drift.—The entire helicopter has a tendency to move in the direction of tail rotor thrust (to the right) when hovering. This movement is often referred to as drift. To counteract this drift, the rotor mast in some helicopters is rigged slightly to the left side so that the tip-path plane has a built-in tilt to the left thus producing a small sideward thrust. In other helicopters, drift is overcome by rigging the cyclic pitch system to give the required amount of tilt to the tip-path plane (fig. 20).

Ground effect.—When a helicopter is in a hovering position close to the ground, the rotor blades will be displacing air downward through the disc faster than it can escape from beneath the helicopter. This builds up a cushion of denser air between the ground and the helicopter (fig. 21). This cushion of denser air, referred to as ground effect, aids in supporting the helicopter while

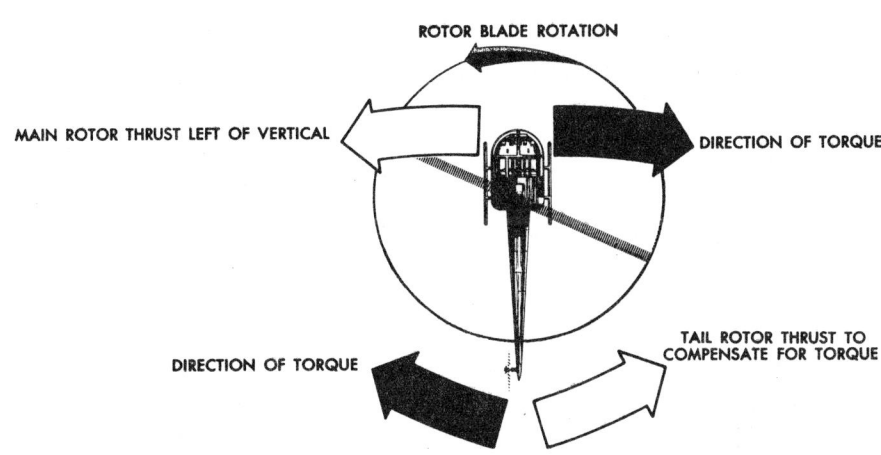

FIGURE 20.—Drift, caused by tail rotor thrust, is compensated for by rigging the mast or cyclic pitch system to have a built-in tilt of the tip-path plane to the left.

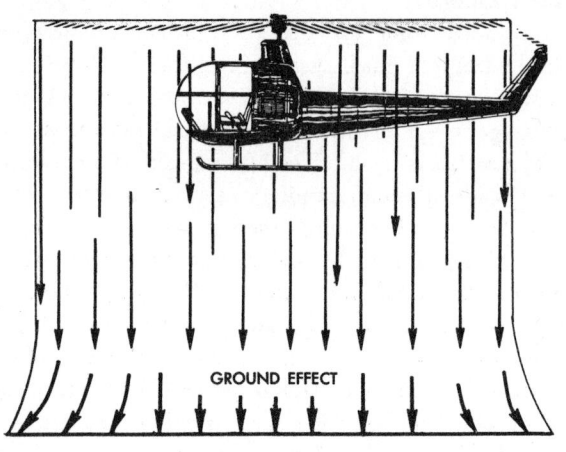

GROUND EFFECT

FIGURE 21.—Ground effect results from the cushion of denser air built up between the ground and helicopter by the air displaced downward by the rotor.

hovering. It is usually effective to a height of approximately one-half the rotor disc diameter. At approximately 3 to 5 miles per hour groundspeed, the helicopter will leave its ground cushion.

Translational lift.—Translational lift is that additional lift obtained when entering horizontal flight, due to the increased efficiency of the rotor system. The rotor system produces more lift in forward flight because the higher inflow velocity supplies the rotor disc with a greater mass of air per unit time upon which to work than it receives while hovering. Translational lift is present with any horizontal movement although the increase will not be noticeable until airspeed reaches approximately 15 miles per hour. The additional lift available at this speed is referred to as "effective translational lift" and is easily recognized in actual flight by the increased performance of the helicopter.

Since translational lift depends upon airspeed rather than groundspeed, the helicopter does not have to be in horizontal flight to be affected. Translational lift will be present during hovering flight in a wind—the amount being proportional to the wind velocity—and effective translational lift will be present when hovering in winds of 15 MPH or more.

Transverse flow effect.—In forward flight, air passing through the rear portion of the rotor disc has a higher downwash velocity than air passing through the forward portion. This is because the air passing through the rear portion has been accelerated for a longer time than the air passing through the forward portion. In other words, the relative wind has a higher velocity at the rear portion of the rotor disc than at the forward portion. This increase in relative wind velocity and resultant increased lift, in combination with gyroscopic precession, causes the rotor disc to tilt to the left side. The lift on the rearward part of the rotor disc is greater than on the forward part. According to the principle of gyroscopic precession, maximum deflection of the rotor blades occurs 90° later in the direction of rotation. This means that the rotor blades will reach maximum upward deflection on the right side and maximum downward deflection on the left side. The overall effect is a tendency of the helicopter to roll to the left. This effect is most noticeable on entry into effective translational lift where it may be accompanied by vibration.

Pendular action.—Since the fuselage of the helicopter is suspended from a single point and has considerable mass, it is free to oscillate either longitudinally or laterally in the same way as a pendulum (fig. 22). This pendular action can be exaggerated by overcontrolling; therefore, control stick movements should be decidedly moderate.

AUTOROTATION

Autorotation is the term used for the flight condition during which no engine power is supplied and the main rotor is driven only by the action of the relative wind. It is the means of safely landing a helicopter after engine failure or certain other emergencies. The helicopter transmission or power train is designed so that the engine, when it stops, is automatically disengaged from the main rotor system to allow the main rotor to rotate freely in its original direction. For obvious reasons, this autorotational capability is not only a most desirable characteristic but is indeed a capability required of all helicopters before FAA certification is granted.

When engine power is being supplied to the main rotor, the flow of air is downward through the rotor. When engine power is not being supplied to the main rotor, that is, when the helicopter is in autorotation, the flow of air is upward through the rotor. It is this upward flow of air that causes the rotor to continue turning after engine failure.

The portion of the rotor blade that produces the forces that cause the rotor to turn when the engine is no longer supplying power to the rotor, is that portion between approximately 25 percent and 70 percent of the radius outward from the center. This portion is often referred to as the "autorotative or driving region" (fig. 23). Aerodynamic forces along this portion of the blade tend to speed up the blade rotation.

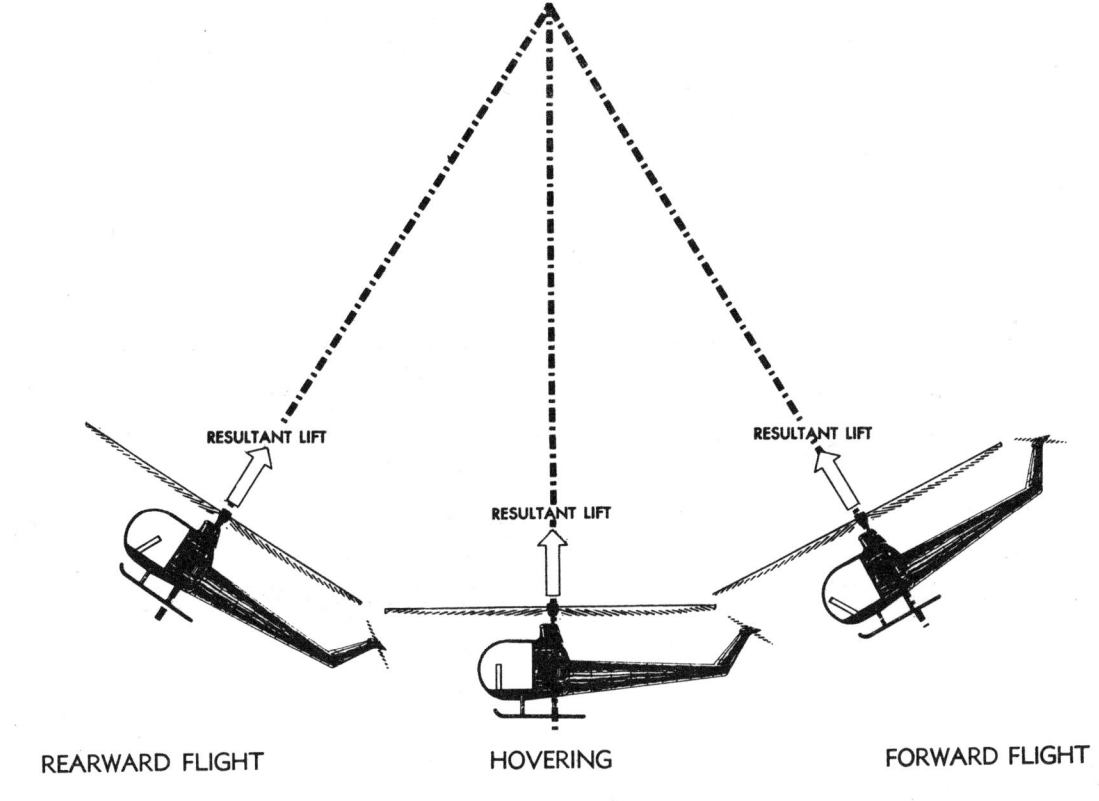

RESULTANT LIFT

RESULTANT LIFT

RESULTANT LIFT

REARWARD FLIGHT HOVERING FORWARD FLIGHT

FIGURE 22.—Since the helicopter is suspended from the rotor mast head, it acts much like a pendulum.

The inner 25 percent of the rotor blade, referred to as the "stall region," operates above its maximum angle of attack (stall angle), thereby contributing little lift but considerable drag which tends to slow the blade rotation.

The outer 30 percent of the rotor blade is known as the "propeller or driven region." Aerodynamic forces here result in a small drag force which tends to slow the tip portion of the blade.

The aerodynamic regions as described above are for vertical autorotations. During forward flight autorotations, these regions are displaced across the rotor disc to the left (fig. 23).

Rotor RPM during autorotation

Rotor RPM stabilizes when the autorotative forces (thrust) of the "driving region" and the antiautorotative forces (drag) of the "driven region" and "stall region" are equal. Assume that rotor RPM has been increased by entering an updraft; a general lessening in angle of attack will follow along the entire blade. This produces a change in aerodynamic force vectors which results ·in an overall decrease in the autorotative forces and the rotor

tends to slow down. If rotor RPM has been decreased by entering a downdraft, autorotative forces will tend to accelerate the rotor back to its equilibrium RPM.

Assuming a constant collective pitch setting, that is, a constant rotor blade pitch angle, an overall greater angle of attack of the rotor disc (as in a flare) increases rotor RPM; a lessening in overall angle of attack (such as "pushing over" into a descent) decreases rotor RPM.

Flares during autorotation

Forward speed during autorotative descent permits a pilot to incline the rotor disc rearward, thus causing a flare. The additional induced lift created by the greater volume of air momentarily checks forward speed as well as descent. The greater volume of air acting on the rotor disc will normally increase rotor RPM during the flare. As the forward speed and descent rate near zero, the upward flow of air has practically ceased and rotor RPM again decreases; the helicopter settles at a slightly increased rate but with reduced forward speed. The flare enables the pilot to make an emergency landing on a definite spot with little or no landing roll or skid.

17

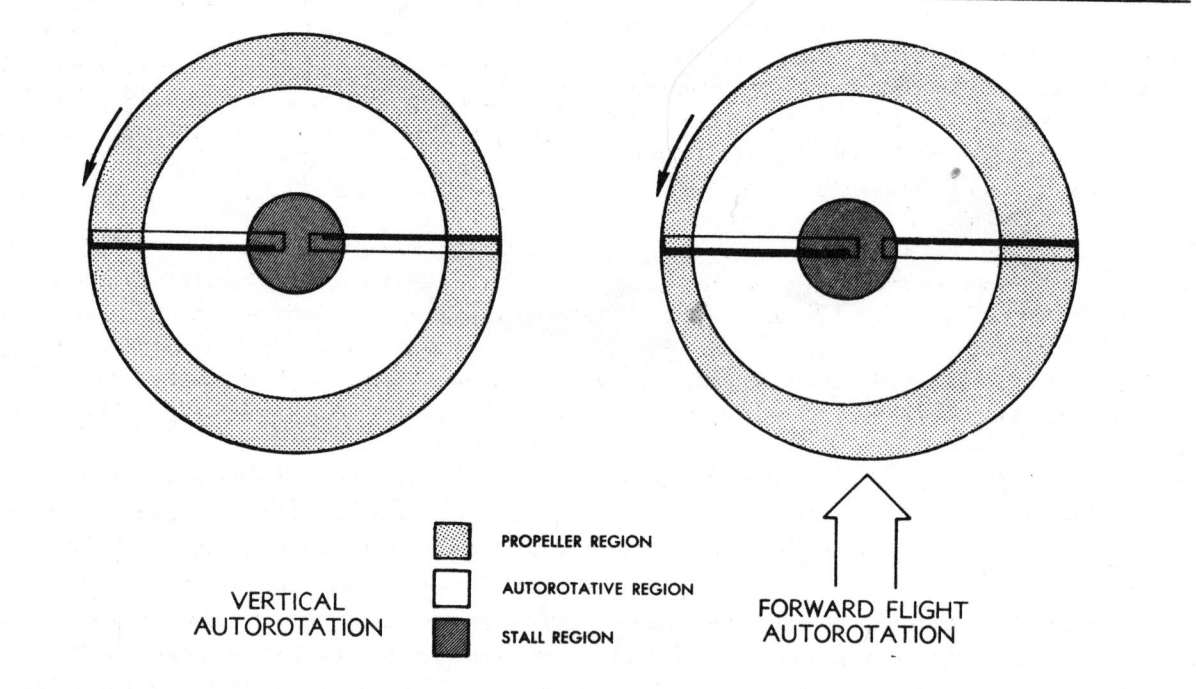

PROPELLER REGION

AUTOROTATIVE REGION

STALL REGION

VERTICAL
AUTOROTATION

FORWARD FLIGHT
AUTOROTATION

FIGURE 23.—Contribution of various portions of the rotor disc to the maintenance of RPM during an autorotation—vertical autorotation (left); forward flight autorotation (right).

Chapter 3. LOADS AND LOAD FACTORS

Before discussing loads and load factors, it is first necessary to discuss the lift forces during turns.

Lift components of a turn

Turns are made in a helicopter, as in an airplane, by banking. In forward flight, the rotor disc is tilted forward which also tilts the total lift–thrust force of the rotor disc forward. This total force is the resultant of a vertical component, lift, and a horizontal component, thrust, acting forward. When the helicopter is placed in a bank, the rotor disc is further tilted sideward. This causes the lift component to be tilted sideward, which in turn, is divided into two components—one acting vertically that opposes weight, the other acting horizontally to the side and opposes centrifugal force (fig. 24). It is this horizontal component of lift that pulls the helicopter in the direction of bank and thus causes it to turn. Briefly then, we can say that *a turn is produced by banking the helicopter thus allowing the lift of the rotor disc to pull the helicopter from its straight course.*

As the angle of bank increases, the total lift force is tilted more toward the horizontal, thus causing the rate of turn to increase because more lift is acting horizontally. Since the resultant lifting force acts more horizontally,

the effect of lift acting vertically (vertical component) is decreased (fig. 25). In order to compensate for this decreased vertical lift, the angle of attack of the rotor blades must be increased in order to maintain altitude. The steeper the angle of bank, the greater the angle of attack of the rotor blades required to maintain altitude. Thus, with an increase in bank and a greater angle of attack, the resultant lifting force will be increased and the rate of turn will be faster.

Loads

Helicopter strength is measured basically by the total load the rotor blades are capable of carrying without permanent damage. The load imposed upon the rotor blades depends largely on the type of flight. The blades must support not only the weight of the helicopter and its contents (gross weight), but also the additional loads imposed during maneuvers.

In straight-and-level flight, the rotor blades support a weight equal to the helicopter and its contents. So long as the helicopter is moving at a constant altitude and airspeed in a straight line, the load on the blades remains constant. When the helicopter assumes a curved flight path—all types of turns (except hovering turns utilizing

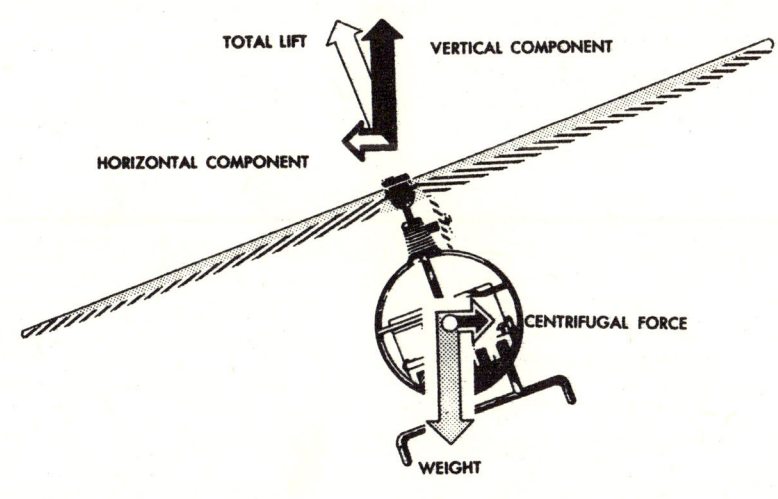

FIGURE 24.—Forces acting on a helicopter in a turn. Lift causes the helicopter to turn when it is banked.

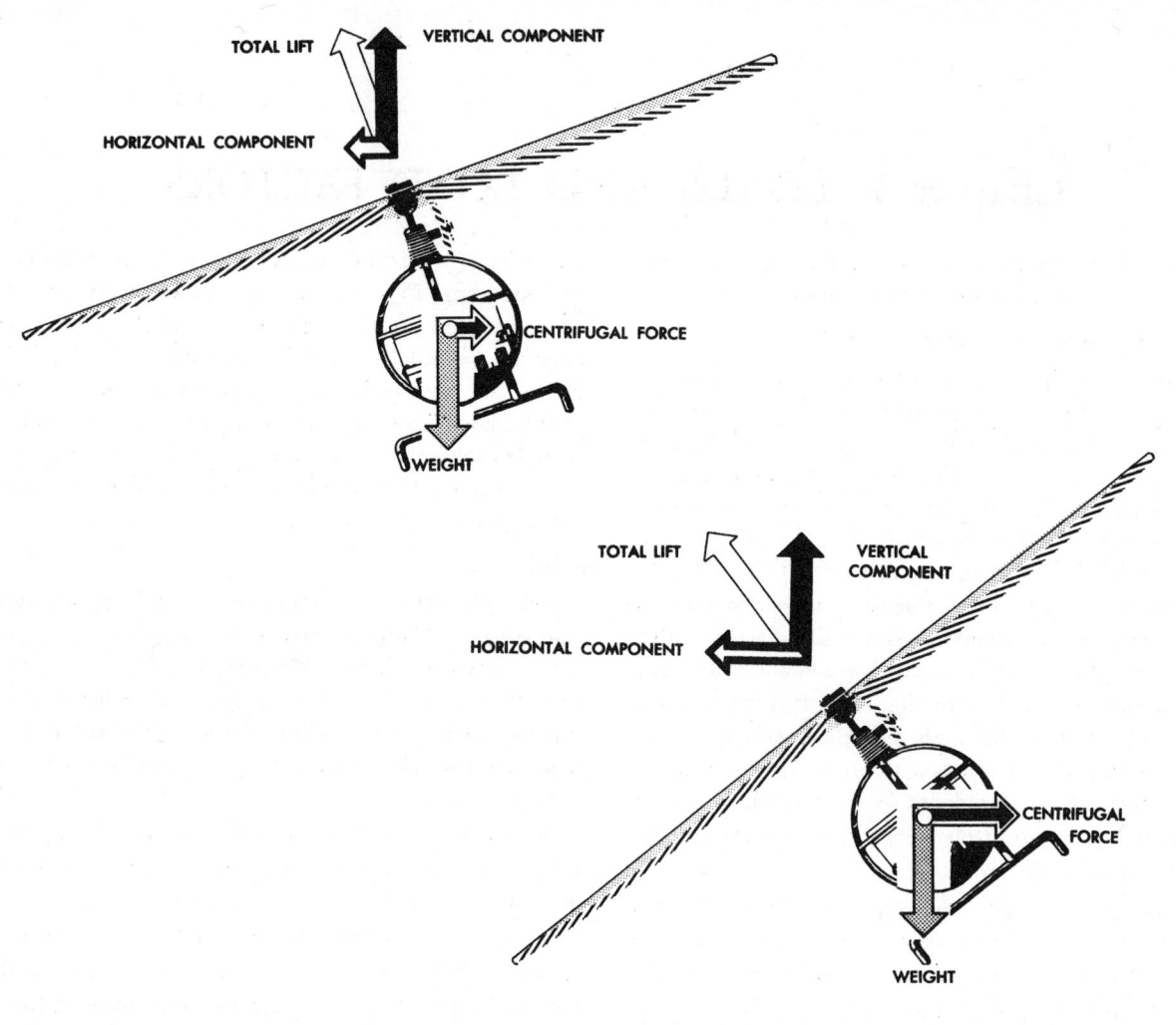

Figure 25.—Relationship between angle of bank and total lift force. As the angle of bank increases, the total lift force is tilted more horizontally, resulting in a faster rate of turn.

pedals only), flares, and pullouts from dives—the actual load on the blades will be much greater because of the centrifugal force produced by the curved flight. This additional load results in the development of much greater stresses on the rotor blades.

Load factor

The load factor is the actual load on the rotor blades at any time, divided by the normal load or gross weight (weight of the helicopter and its contents). Any time a helicopter flies in a curved flight path, the load supported by the rotor blades is greater than the total weight of the helicopter. The tighter the curved flight path, that is, the steeper the bank, or the more rapid the flare or pull-out from a dive, the greater the load supported by the rotor; therefore, the greater the load factor.

The load factor and, hence, apparent gross weight increase is relatively small in banks up to 30° (fig. 26). Even so, under the right set of adverse circumstances, such as high-density altitude, gusty air, high gross weight, and poor pilot technique, sufficient power may not be available to maintain altitude and airspeed. Above 30° of bank, the apparent increase in gross weight soars. At 30° of bank, the apparent increase is only 16 percent, but at 60°, it is 100 percent (fig. 26).

If the weight of the helicopter is 1,600 pounds, the weight supported by the rotor in a 30° bank at a constant altitude would be 1,856 pounds (1,600 + 256). In a 60° bank, it would be 3,200 pounds; and in an 80° bank, it would be almost six times as much, or 8,000 pounds.

One additional cause of large load factors is rough or turbulent air. The severe vertical gusts produced by turbulence can cause a sudden increase in angle of attack, resulting in increased rotor blade loads that are resisted by the inertia of the helicopter.

In order to be certificated by FAA, each helicopter must have a maximum permissible limit load factor that should not be exceeded. As a pilot, you should have the basic information necessary to fly a helicopter safely within its structural limitations. Be familiar with the situations in which the load factor may approach maximum, and avoid them. If you meet such situations inadvertently, you must know the proper technique.

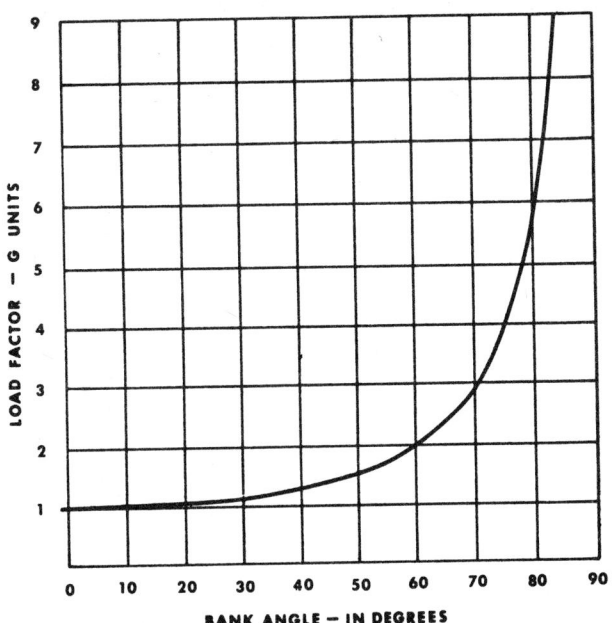

FIGURE 26.—Load factor chart.

Chapter 4. FUNCTION OF THE CONTROLS

There are four controls in the helicopter that the pilot must use during flight (fig. 27). They are (1) collective pitch control; (2) throttle control; (3) anti-torque pedals (auxiliary or tail rotor control); and (4) cyclic pitch control.

Collective pitch control

The collective pitch lever or stick is located by the left side of the pilot's seat and is operated with his left hand (fig. 28). This lever moves up and down pivoting about the aft end and, through a series of mechanical linkages, changes the pitch angle of the main rotor blades. As the collective pitch lever is raised, there is a simultaneous and equal increase in the pitch angle of all the main rotor blades; as the lever is lowered, there is a simultaneous and equal decrease in the pitch angle. The amount of movement of the lever determines the amount of blade pitch change.

As the pitch angle of the rotor blades is changed, the angle of attack of each blade will also be changed. A change in the angle of attack changes the drag on the rotor blades. As the angle of attack increases, drag increases and the rotor RPM and the engine RPM (the needles are joined) tend to decrease; as the angle of attack decreases, drag decreases and the RPM tends to increase. Since it is essential that the RPM remain constant, there must be some means of making a proportionate change in power to compensate for the change in drag. This coordination of power change with blade pitch angle change is controlled through a collective pitch lever-throttle control cam linkage which automatically increases power when the collective pitch lever is raised and decreases power when the lever is lowered.

The collective pitch control is the primary altitude control. Raising the collective pitch lever increases the rotor's lift and, through the cam linkage with the throttle, increases engine power. The collective pitch control is, therefore, the primary manifold pressure control (fig. 29).

Throttle control

The throttle is mounted on the forward end of the collective pitch lever in the form of a motorcycle-type twist grip. The function of the throttle is to regulate RPM.

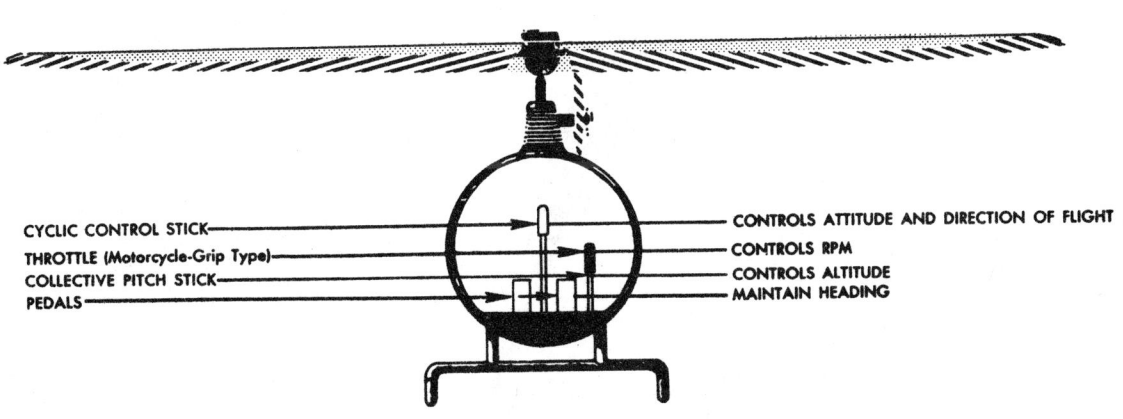

CYCLIC CONTROL STICK ———————————— CONTROLS ATTITUDE AND DIRECTION OF FLIGHT
THROTTLE (Motorcycle-Grip Type) ——————— CONTROLS RPM
COLLECTIVE PITCH STICK ———————————— CONTROLS ALTITUDE
PEDALS ————————————————————— MAINTAIN HEADING

FIGURE 27.—Controls of the helicopter and the principal function of each control.

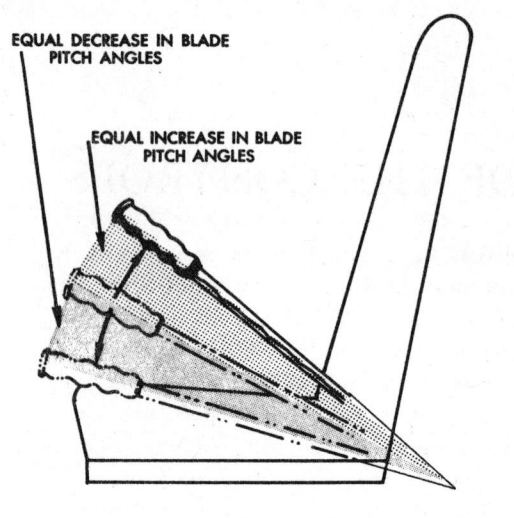

EQUAL DECREASE IN BLADE PITCH ANGLES

EQUAL INCREASE IN BLADE PITCH ANGLES

FIGURE 28.—Collective pitch stick movement produces equal changes in blade pitch angles.

If the collective pitch-throttle synchronization unit does not automatically maintain a constant RPM when a change is made in the collective pitch stick position, the throttle may be moved manually with the twist grip to make further adjustments of engine RPM. Twisting the throttle outboard increases RPM; twisting it inboard decreases RPM (fig. 30).

The throttle must be coordinated with the collective pitch so that a correct rotor RPM is maintained. The throttle, therefore, is the primary RPM control (fig. 29).

Collective pitch–throttle coordination

Collective pitch is the primary control for manifold pressure; the throttle is the primary control for RPM. Since the collective pitch control also influences RPM and the throttle also influences manifold pressure, each is considered to be a secondary control of the other's functions. Therefore, the pilot must analyze both his tachometer (RPM indicator) and his manifold pressure gage to determine which control to use and how much. To best illustrate the relationship, a few problems with solutions follow:

> Problem: RPM low, manifold pressure low.
> Solution: Increasing the throttle will increase the RPM *and* the manifold pressure.
> Problem: RPM low, manifold pressure high.
> Solution: Lowering the collective pitch will reduce the manifold pressure, decrease drag on the rotor, and therefore, increase the RPM.
> Problem: RPM high, manifold pressure high.
> Solution: Decreasing the throttle reduces the RPM *and* the manifold pressure.
> Problem: RPM high, manifold pressure low.
> Solution: Raising the collective pitch will increase the manifold pressure, increase drag on the rotor, and therefore, decrease the RPM.

These problems illustrate how one control change accomplishes *two* purposes. An extension of the reasoning used in the solutions will show how various combinations of control inputs can be coordinated to achieve any desired

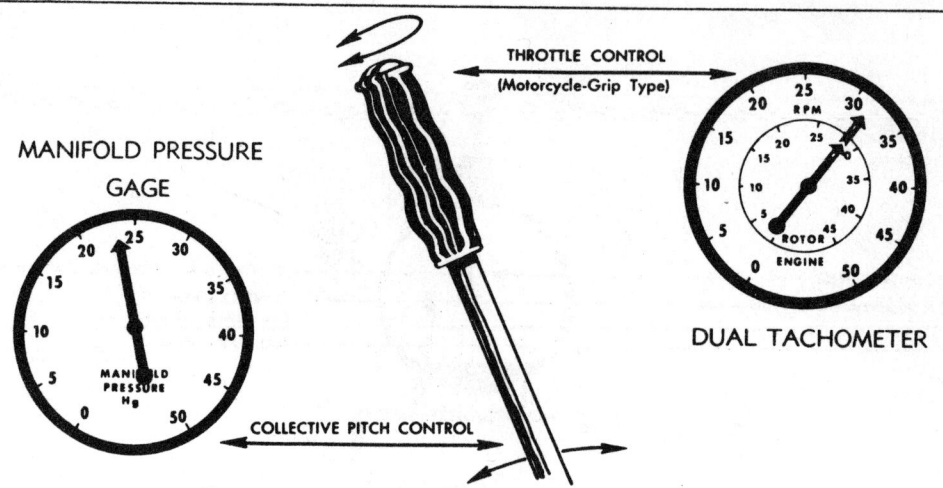

FIGURE 29.—Collective pitch stick is primary control for manifold pressure; throttle is primary control for RPM. A change in either control results in a change in both manifold pressure and RPM.

RPM-manifold pressure setting. As with any other aircraft controls, large adjustments of either collective pitch or throttle should be avoided. All corrections should be accomplished through the use of smooth pressures.

Antitorque pedals

The thrust produced by the auxiliary (tail) rotor is governed by the position of the antitorque pedals. These pedals are located as shown in figure 27. They are linked to a pitch change mechanism in the tail rotor gear box to permit the pilot to increase or decrease the pitch of the tail rotor blades. The primary purpose of the tail rotor and its controls is to counteract the torque effect of the main rotor.

Heading control

The tail rotor and its controls not only enable the pilot to counteract the torque of the main rotor during flight, but also enable him to control the heading of the helicopter during hovering flight, hovering turns, and hovering patterns. It should be thoroughly understood that in forward flight, the pedals are not used to control the heading of the helicopter (except during portions of crosswind takeoffs and approaches); rather, they are used to compensate for torque to put the helicopter in longitudinal trim so that coordinated flight (that is, neither slipping nor skidding) can be maintained. The cyclic control is used to change heading by making a coordinated turn to the desired direction.

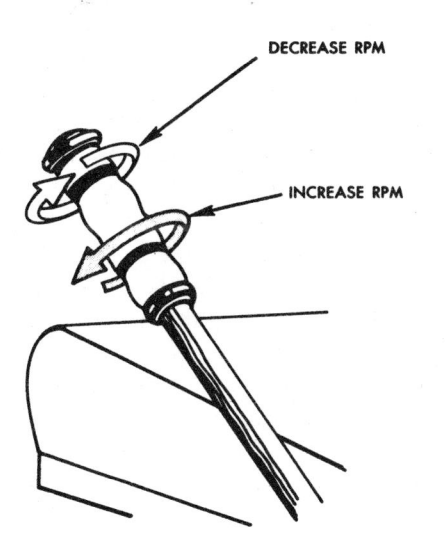

DECREASE RPM

INCREASE RPM

FIGURE 30.—Throttle control: Rotating the throttle outboard (viewed from the top) increases RPM; rotating it inboard decreases RPM.

The thrust of the tail rotor is dependent upon the pitch angle of the tail rotor blades and, to a certain extent, upon the main rotor RPM. (For this particular discussion, we will assume that the main rotor RPM remains constant.) The pitch angle of the tail rotor blades determines the size of the bite of air the blades take as they rotate. The tail rotor may have a positive pitch angle, that is, the rotor bites the air to the right which tends to pull the tail to the right; or it may have a negative pitch angle in which case the rotor bites the air to the left, tending to pull the tail to the left; or it may have zero pitch, in which case it produces no thrust in either direction.

With the right pedal moved forward, the tail rotor either has a negative pitch angle or a small positive pitch angle—the farther forward the right pedal is, the larger the negative pitch angle; the nearer the right pedal is to the neutral position, the more positive pitch angle the tail rotor will have; and somewhere in between, the tail rotor will have a zero pitch angle. As the left pedal is moved forward of the neutral position, the positive pitch angle of the tail rotor increases until it becomes maximum with full forward displacement of the left pedal.

With a negative pitch angle, the tail rotor thrust is working in the same direction as torque reaction of the main rotor. With a small positive pitch angle, the tail rotor does not produce sufficient thrust to overcome the torque effect of the main rotor during cruising flight. Therefore, if the right pedal is displaced forward of neutral during cruising flight, the tail rotor thrust will not overcome the torque effect and the nose will yaw to the right (fig. 31 left).

With the pedals in the neutral position, the tail rotor has a medium positive pitch angle. In medium positive pitch, the tail rotor thrust approximately equals the torque of the main rotor during cruising flight, so the helicopter will maintain a constant heading in level flight (fig. 31 center).

With the left pedal in a forward position, the tail rotor is in a high positive pitch position. In a high positive pitch position, tail rotor thrust exceeds the thrust needed to overcome torque effect during cruising flight so the helicopter nose will yaw to the left (fig. 31 right).

The above explanation is based on cruising power and airspeed. Since the amount of torque is dependent on the amount of engine power being supplied to the main rotor, the relative positions of the pedals required to counteract torque will depend upon the amount of power being used at any time. In general, however, the less power being used, the greater the forward displacement of the right

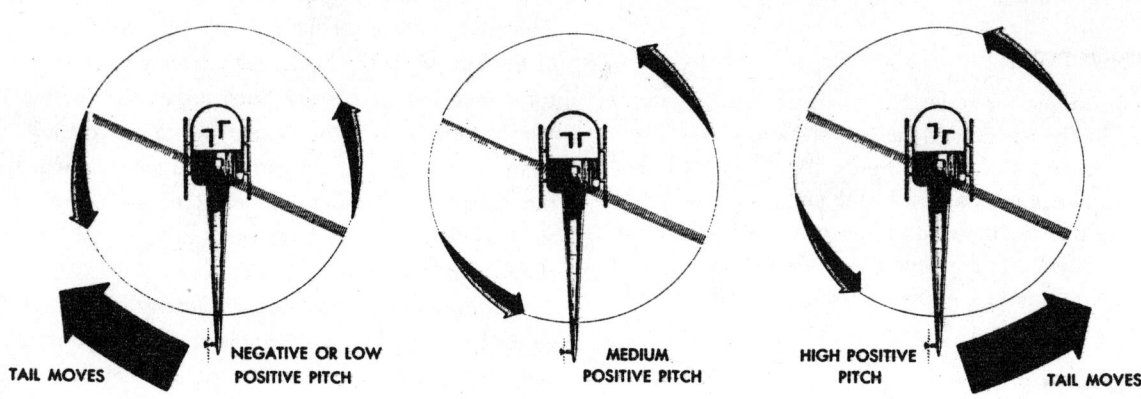

TAIL MOVES — NEGATIVE OR LOW POSITIVE PITCH — MEDIUM POSITIVE PITCH — HIGH POSITIVE PITCH — **TAIL MOVES**

FIGURE 31.—Tail rotor pitch angle and thrust in relation to pedal positions during cruising flight.

pedal that is required; the greater the power being used, the greater the forward displacement of the left pedal.

The maximum positive pitch angle of the tail rotor is generally somewhat greater than the maximum negative pitch angle available. This is because the primary purpose of the tail rotor is to counteract the torque of the main rotor. The capability for tail rotors to produce thrust to the left (negative pitch angle) is necessary because, during autorotation, the drag of the transmission tends to yaw the nose to the left—in the same direction that the main rotor is turning.

Cyclic pitch control

As discussed previously, the total lift–thrust force is always perpendicular to the tip-path plane of the main rotor. When the tip-path plane is tilted away from the horizontal, the lift–thrust force is divided into two components—the horizontal acting force, thrust; and the upward acting force, lift (fig. 11). *The purpose of the cyclic pitch control is to tilt the tip-path plane in the direction that horizontal movement is desired.* The thrust component then pulls the helicopter in the direction of rotor tilt. The cyclic control has no effect on the magnitude of the total lift–thrust force, but merely changes the direction of this force, thus controlling the attitude and airspeed of the helicopter.

The rotor disc tilts in the direction that pressure is applied to the cyclic. If the cyclic stick is moved forward, the rotor disc tilts forward; if the cyclic is moved aft, the rotor disc tilts aft, and so on (fig. 32).

So that the rotor disc will always tilt in the direction that the cyclic stick is displaced, the mechanical linkage between the cyclic stick and the rotor (through the swash plate) must be such that the maximum downward deflection of the blades is reached in the direction the stick is displaced and maximum upward deflection is reached in the opposite direction. Otherwise, the pilot would have a difficult job of relating the direction of cyclic stick displacement to the rotor disc tilt. This is accomplished through the mechanical linkage which decreases the pitch angle of the rotor blades 90° before they reach the direction of displacement of the cyclic stick and increases the pitch angle of the rotor blades 90° after they pass the direction of displacement of the cyclic stick. Any increase in pitch angle increases the angle of attack; any decrease in pitch angle decreases the angle of attack.

For example, as the cyclic stick is displaced forward, the angle of attack is decreased as the rotor blades pass the 90° position to the pilot's right and is increased as the blades pass the 90° position to the pilot's left. Because of gyroscopic precession, maximum downward deflection of the rotor blades is forward and maximum upward deflection is aft, causing the rotor disc to tilt forward in the same direction as cyclic stick displacement. A similar analysis could be made for any direction of displacement of the cyclic stick.

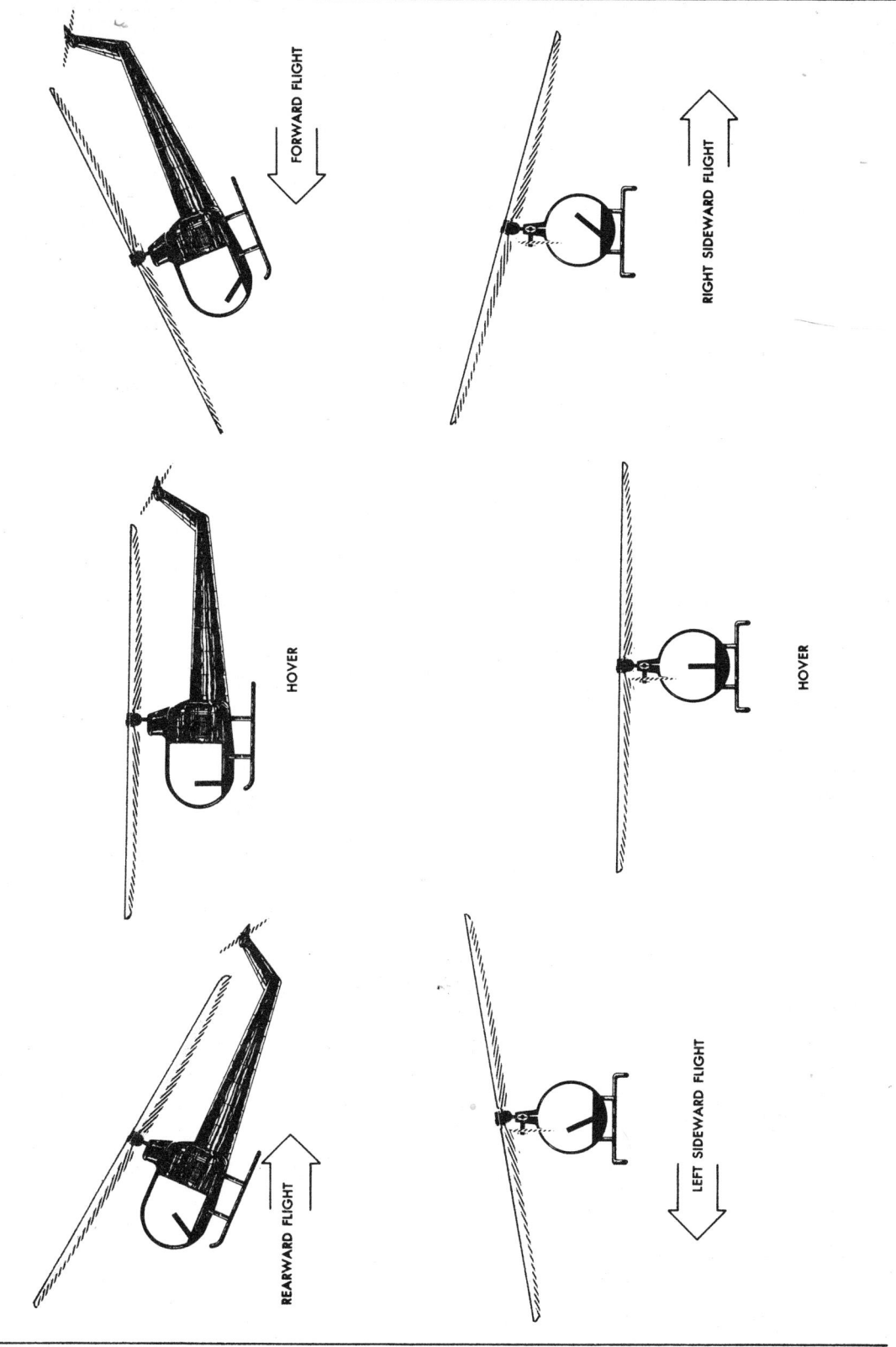

FORWARD FLIGHT

RIGHT SIDEWARD FLIGHT

HOVER

HOVER

REARWARD FLIGHT

LEFT SIDEWARD FLIGHT

FIGURE 32.—Relationship of cyclic stick position to rotor disc position and helicopter movement.

Chapter 5. OTHER HELICOPTER COMPONENTS AND THEIR FUNCTIONS

In the preceding chapter, the control system and its functions were discussed in detail. In this chapter, some of the other components and their functions will be discussed briefly to give the reader some familiarity with the aircraft he will be flying.

TRANSMISSION SYSTEM

The transmission system transmits engine power to the main rotor, tail rotor, generator, and other accessories.

The engine of a helicopter must operate at a relatively high speed while the main rotor turns at a much lower speed. This speed reduction is accomplished through reduction gears in the transmission system and is generally somewhere between 6 to 1 and 9 to 1 (that is, between 6 and 9 engine RPM's to 1 main rotor RPM). In a helicopter with a 6 to 1 ratio, if the engine turns at 2700 RPM, the main rotor turns at 450 RPM. With a 9 to 1 ratio, if the engine turns at 2700 RPM, the main rotor turns at 300 RPM. When the rotor tachometer needle and the engine tachometer needle are superimposed over each other (fig. 29), the ratio of the engine RPM to the rotor RPM is the same as the gear reduction ratio.

CLUTCH

In the conventional airplane, it is standard practice to have the engine and the propeller permanently connected. The propeller serves as a flywheel; there is no reason for the propeller to be at a standstill when the engine is running. In the helicopter, there is a different relation between the engine and rotor.

Because of the much greater weight of a helicopter rotor in relation to the power of the engine than the weight of a propeller in relation to the power of the engine in an airplane, it is necessary to have the rotor disconnected from the engine to relieve the starter load. For this reason, it is necessary to have a clutch between the engine and rotor. The clutch allows the engine to be started and gradually assume the load of driving the heavy rotor system.

The clutch does not provide disengagement of the engine from the rotor system for autorotation. This is provided through another device.

Centrifugal clutch

In this type of clutch, contact between the inner and outer parts of the clutch is made by the spring-loaded clutch shoes. The inner portion of the clutch, the clutch shoes, is rotated by the engine; the outer portion of the clutch, the clutch drum, is connected to the main rotor through the transmission. At low engine speeds, the clutch shoes are held out of contact with the clutch drum by the springs. As engine speed increases, centrifugal force throws the clutch shoes outward until they contact the clutch drum and motion is transmitted from the engine drive shaft to the input drive shaft of the transmission. The rotor starts to turn, slowly at first, but with increasing speed as the friction between the clutch shoes and drum increases. Slippage of the clutch will be experienced until this friction develops sufficiently to drive the drum at engine RPM. As the clutch becomes fully engaged, the rotor system will be driven at the equivalent of engine RPM and the rotor tachometer needle and engine tachometer needle will join or "marry," that is, one needle will be superimposed over the other.

The rotor RPM equivalent to the engine RPM depends upon the gear reduction ratio between the engine and rotor system for the particular helicopter. (See Transmission System.)

Friction or belt drive system clutch

This type of clutch is manually engaged by the pilot through a lever in the cockpit. Power from the engine drive shaft is transmitted to the transmission drive shaft by a series of friction discs or belts. With this type of clutch, it is possible to start the engine and warm it up without engaging the rotor.

FREEWHEELING UNIT

The freewheeling coupling provides for autorotative capabilities by automatically disconnecting the rotor system from the engine when the engine stops or slows below the equivalent of rotor RPM. When the engine is disconnected from the rotor system through the automatic action of the freewheeling coupling, the transmission continues to rotate with the main rotor thereby enabling

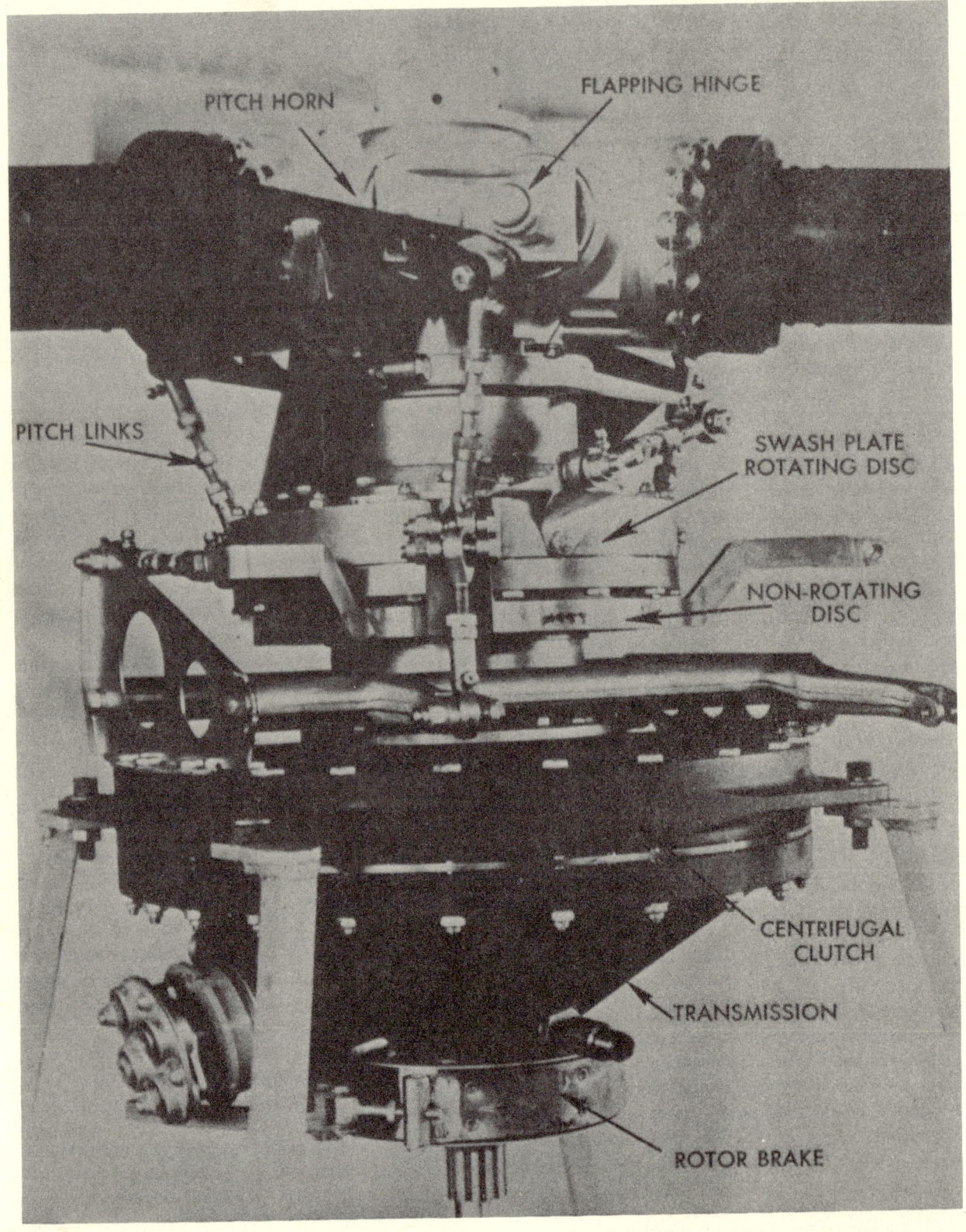

PITCH HORN

FLAPPING HINGE

PITCH LINKS

SWASH PLATE ROTATING DISC

NON-ROTATING DISC

CENTRIFUGAL CLUTCH

TRANSMISSION

ROTOR BRAKE

Courtesy Brantly Helicopter Corporation

FIGURE 33.—Various components of the rotor system in one helicopter.

the tail rotor to continue turning at its normal rate. This permits the pilot to maintain directional control during autorotation.

SWASH PLATE ASSEMBLY

The swash plate consists of two primary elements through which the rotor mast passes (figs. 33, 36, and 37). One element is a disc, linked to the cyclic pitch control. This disc is capable of tilting in any direction but does not rotate as the rotor rotates. This nonrotating disc, often referred to as the "stationary star", is attached by a bearing surface to a second disc, often referred to as the "rotating star," which turns with the rotor and is mechanically linked to the rotor blade pitch horns.

The rotor blade pitch horns are placed approximately 90° ahead of or behind the blade on which they control the pitch change (figs. 33 and 37). If this were not done, gyroscopic precession would cause the movement of the helicopter to be 90° out of phase with the movement of the cyclic pitch stick, that is, if the cyclic stick were displaced to the right, the helicopter would move forward; if the cyclic stick were displaced forward, the helicopter would move to the left, and so on.

The illustration in figure 34 shows the pitch horns 90° ahead of the blade in the plane of rotation. Figure 37 shows them 90° behind. Whether they are ahead of or behind the blade will depend on the mechanical linkage arrangement between the cyclic stick, swash plate, and pitch horns. It might help to understand the relationship between cyclic stick movement and blade pitch change if the relationship between cyclic stick movement and the rotor blade pitch horn is understood. If the pitch horn is 90° ahead of the blade, blade pitch decrease takes place as the pitch horn passes the direction in which the cyclic

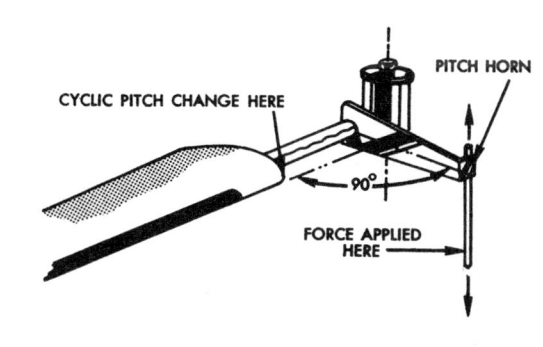

FIGURE 34.—Rotor blade pitch horns are located 90° ahead of or behind (depending on the manufacturer) the rotor blade so that helicopter reaction will be in the direction of cyclic stick displacement.

stick is displaced. Blade pitch increase takes place as the pitch horn passes the direction opposite to the displacement. If the pitch horn is 90° behind the blade, blade pitch decrease takes place as the pitch horn passes the direction opposite to the displacement of the cyclic stick. Blade pitch increase takes place as the pitch horn passes the direction of displacement. In either case, however, blade pitch decrease takes place 90° ahead of cyclic stick position and blade pitch increase takes place 90° after passing cyclic stick position. Thus, maximum downward deflection of the rotor blades occurs in the same direction as cyclic stick displacement and maximum upward deflection occurs in the opposite direction.

As an example, when the cyclic stick is displaced forward, the swash plate nonrotating disc tilts forward and the swash plate rotating disc follows this forward tilt (fig. 35). Since the mechanical linkage from the rotating disc to the rotor blade pitch horns is 90° ahead of or behind the cyclic pitch change, the pitch angle is decreased as the rotor blades pass 90° to the pilot's right and increased as the rotor blades pass 90° to the pilot's left. Because of gyroscopic precession, maximum blade deflection occurs 90° later in the cycle of rotation. Thus, maximum downward deflection of the rotor blades is forward (same direction as cyclic stick displacement) and maximum upward deflection is aft, causing the rotor disc to tilt forward in the same direction as cyclic stick displacement.

MAIN ROTOR SYSTEM

There are three fundamental types of main rotor systems: fully articulated rotors, semirigid rotors, and rigid rotors.

Fully articulated rotor systems

Fully articulated rotor systems generally consist of three or more rotor blades. In a fully articulated rotor system each rotor blade is attached to the rotor hub by a horizontal hinge, called the flapping hinge, which permits the blades to flap up and down. Each blade can move up and down independently of the others. The flapping hinge may be located at varying distances from the rotor hub, and there may be more than one. The position is chosen by each manufacturer primarily with regard to stability and control.

Each rotor blade is also attached to the hub by a vertical hinge, called a drag or lag hinge, that permits each blade, independently of the others, to move back and forth in the plane of the rotor disc. This movement is called dragging, lead-lag, or hunting. The location of this hinge is chosen primarily with regard to controlling vibration. Dampers are normally incorporated in the design of this type rotor system to prevent excessive motion about the drag hinge.

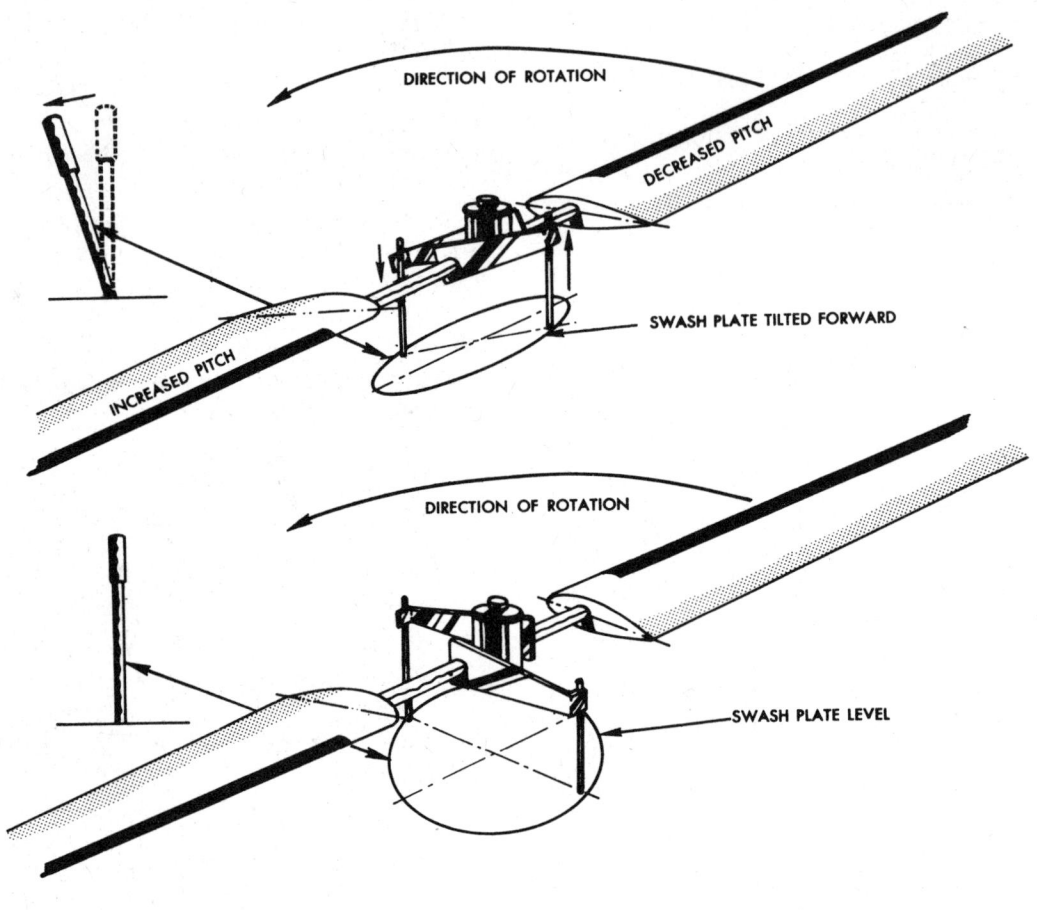

FIGURE 35.—Cyclic stick movements are transmitted by a mechanical linkage through the swash plate to the rotor pitch horns and result in a change in the pitch angle of each rotor blade.

The purpose of the drag hinge and dampers is to absorb the acceleration and deceleration of the rotor blades caused by coriolis effect. Figure 37 points out the flapping hinges and drag hinges.

The blades of a fully articulated rotor can also be feathered, that is, rotated about their spanwise axis. To put it more simply, feathering means the automatic and periodic changing of the pitch angle of the rotor blades.

Summarizing then, each blade of a fully articulated rotor system can flap, drag, and feather independently of the other blades.

Semirigid rotor systems

In a semirigid rotor system, the rotor blades are rigidly interconnected to the hub, but the hub is free to tilt and rock with respect to the rotor shaft. In this system, only two-bladed rotors are used. The rotor flaps as a unit, that is, as one blade flaps up, the other blade flaps down an equal amount.

The hinge which permits the flapping or seesaw effect is called a teetering hinge (fig. 38). The rocking hinge is perpendicular to the teetering hinge and parallel to the rotor blades. This hinge allows the head to rock in response to tilting of the swash plate by cyclic pitch control, thus changing the pitch angle an equal amount on each blade—decreasing it on one and increasing it on the other.

The rotor blades of a semirigid rotor system may or may not require drag hinges depending on whether the system is "underslung." In an underslung system, the rotor blades lie in a plane below the plane containing the rotor hub pivot point. Because of coning, normal rotor

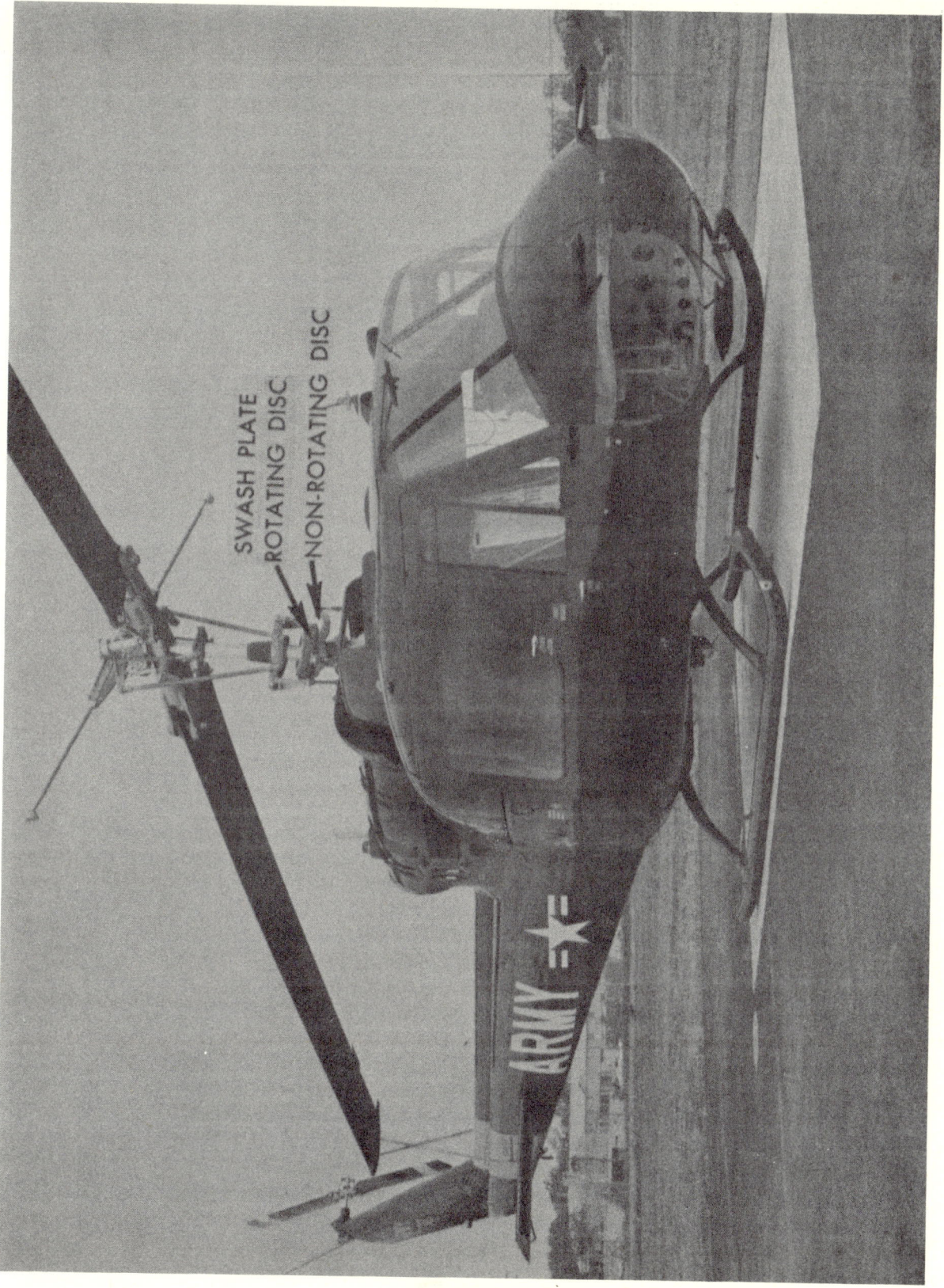

SWASH PLATE
ROTATING DISC
NON-ROTATING DISC

Courtesy Bell Helicopter Corporation

FIGURE 36.—Swash plate system for one helicopter.

33

FLAPPING HINGE

SWASH PLATE
ROTATING DISC

NON ROTATING DISC

LEAD-LAG (DRAG) HINGE

PITCH HORN

Courtesy Hughes Tool Company, Aircraft Division

FIGURE 37.—Various components of the rotor system in one helicopter.

34

TEETERING HINGE ROCKING HINGE TEETERING HINGE

Courtesy Bell Helicopter Corporation

FIGURE 38.—Teetering and rocking hinges on a two-bladed rotor system.

operating RPM will place the center of mass of the rotor blades in approximately the same plane as the rotor hub pivot point. Consequently, the distance of the center of mass from the axis of rotation varies very little. Drag hinges are not needed since the hunting action can be absorbed through blade bending and the movement of the gimbal in the underslung system.

Collective pitch control changes the pitch of each blade simultaneously and an equal amount, either increasing the pitch of both or decreasing the pitch of both.

Summarizing, a semirigid rotor system can flap and feather as a unit.

Rigid rotor systems

In a rigid rotor system the blades, hub, and mast are rigid with respect to each other. In this system, the blades cannot flap or drag but can be feathered.

Extensive research is being done in this area and, at the time of this writing, two makes of rigid rotor helicopters have received FAA certification.

Chapter 6. INTRODUCTION TO THE HELICOPTER FLIGHT MANUAL

It is the responsibility of each pilot to know all pertinent information for each helicopter that he flies. The helicopter flight manual is designed to provide pilots with a general knowledge of the particular helicopter and the information necessary for safe and efficient operation. Its function is not to teach a pilot to fly, but rather to provide the best possible operating instructions under most circumstances. It is not intended as a substitute for sound judgment; emergencies or other unforeseen situations may require modification of the procedures. *A thorough understanding of the contents of the helicopter flight manual will enable pilots to complete flights with maximum efficiency and safety.*

A helicopter flight manual accompanies each certificated helicopter. Although the manual for a particular helicopter may contain much information identical to that contained in the flight manual for other helicopters of the same make and model, it may also contain information which is peculiar only to that helicopter, especially weight and balance information. Helicopter flight manuals are prepared and furnished by the manufacturers. Much of the information in them is required by Federal Aviation Regulation, Part 27, Airworthiness Standards: Normal Category Rotorcraft. However, manufacturers often include additional information that is helpful to the pilot but which is not required.

When the helicopter flight manual contains information required by regulations that does not appear as placards in the helicopter, the manual must be carried in the helicopter at all times. The statement: "This document must be carried in the aircraft at all times" will appear somewhere on the manual if such conditions exist.

The information required by regulations to be included in the helicopter flight manual is generally listed as follows under chapters, sections, headings, or some similar breakdown:

Operating Limitations
Operating Procedures
Performance Information

Operating limitations

All important operating limitations that must be observed during normal operations are covered in this portion of the flight manual. This includes airspeed and rotor limitations, powerplant limitations, weight and loading distribution, flight crew, type of operation, and unusable fuel if the unusable supply in any tank exceeds 1 gallon or 5 percent of the tank capacity. A brief discussion of each of these, along with actual examples excerpted from various FAA-approved helicopter flight manuals follows.

Airspeed limitations.—Required information includes those limiting airspeeds which must be shown on the airspeed indicator by a color coding or must be displayed in the form of a placard. A red radial line must be placed on the airspeed indicator to show the airspeed limit beyond which operation is dangerous (never-exceed speed V_{ne}); a yellow arc is used to indicate cautionary operating ranges; and a green arc is used to indicate safe or normal operating ranges. Here are excerpts from this portion of various helicopter flight manuals:

V_{ne} 100 mph from sea level to 2000 feet. Above 2000 feet, decrease V_{ne} 3 mph per 1000 feet.

* * *

V_{ne} 105 mph sea level to 6000 feet. Above 6000 feet decrease V_{ne} 5 mph per 1000 feet.

This information is sometimes given in the form of a chart (fig. 39).

Rotor limitations.—Required information in this section of the manual includes limiting rotor RPM's and cautionary ranges. These limitations are marked on the tachometer by red radial lines and yellow arcs respectively, with normal operating ranges marked with a green arc. Here are excerpts from two helicopter flight manuals:

Maximum 370 RPM; minimum 333 RPM.

* * *

Maximum 500 RPM for 248–40, –46, –53 blades.
Maximum 472 RPM for 248–100, and –101 blades.
Minimum 400 RPM.

In the latter case, the pilot must know the type of rotor blade used on his particular helicopter.

Powerplant limitations.—Information contained in this portion must explain all powerplant limitations and the required markings on the powerplant instruments. This will include such items as fuel octane rating, idling RPM,

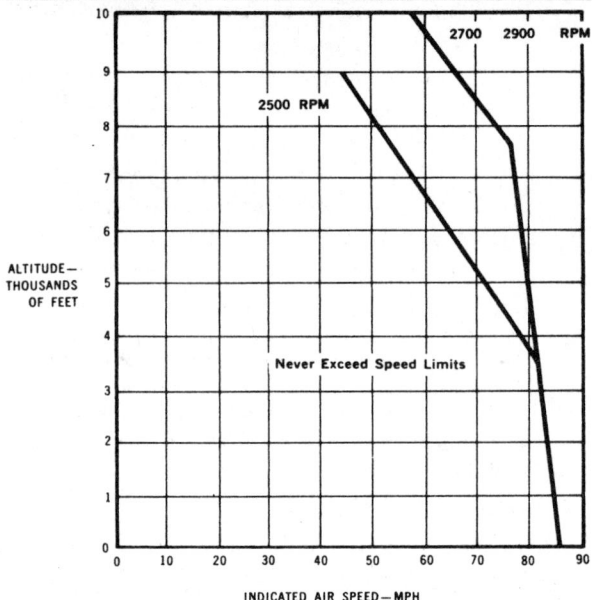

Courtesy Hughes Tool Company, Aircraft Division

FIGURE 39.—Chart showing never-exceed (V_{ne}) speed limits.

operating RPM, manifold pressure, oil pressure, oil temperature, cylinder head temperature, fuel pressure, mixture control and others. See figure 40 for information contained in one flight manual concerning instrument markings.

Weight and loading distribution.—This section must include rotorcraft weights and center-of-gravity limits, together with the items of equipment on which the empty weight is based. This will generally require the inclusion of a chart or graph such as that illustrated in figure 41 from which the pilot can compute the center-of-gravity position for any given loading situation. (See pages 45 and 46 for instructions on the use of this chart.)

Flight crew.—When a flight crew of more than one is required, the number and functions of the minimum flight crew will be described.

Type of operation.—Examples of statements appearing in helicopter flight manuals for light helicopters are:

Basic configuration of the helicopter permits its use as a two-place aircraft. The basic helicopter is approved for VFR operations.

* * *

Basic configuration of the helicopter permits its use as a three-place aircraft.

Unusable fuel.—If the unusable fuel supply in any tank exceeds 1 gallon or 5 percent of the tank capacity, whichever is the greater, a warning shall be provided to indicate to the flight personnel that when the quantity indicator reads zero the remaining fuel in the tank cannot be used for flight.

Operating procedures

This section of the manual contains information concerning normal and emergency procedures, takeoff and landing procedures, appropriate airspeeds peculiar to the rotorcraft's operating characteristics, and other pertinent information necessary for safe operation. This section may include, but not necessarily all, the following procedures: a preflight checklist; before starting engine, starting, warmup, and shutdown procedures; inflight procedures; and such emergency procedures as engine failure, tail rotor failure, hydraulic boost failure, ditching with and without power, and others. The following is an excerpt from one flight manual on the emergency procedure to use in case of engine failure.

TABLE OF INSTRUMENT MARKINGS		
Rotor Tachometer	Yellow Arc	200 to 230 RPM
	Red Line	333 RPM
	Green Arc	333 to 370 RPM
	Red Line	370 RPM
Engine Tachometer	Red Line	3000 RPM
	Green Arc	3000 to 3200 RPM
	Red Line	3200 RPM
Airspeed Indicator	Red Line	105 MPH
Manifold Pressure Gage	Yellow Arc	24.0 to 27.4 in. Hg.
	Red Line	27.4 in. Hg.
Oil Temp Engine	Red Line	40°C
	Green Arc	40° to 107°C
	Red Line	107°C
Oil Temp Transmission	Green Arc	40° to 130°C
	Red Line	130°C
Cylinder Head Temp Gage	Red Line	100°C
	Green Arc	100° to 246°C
	Red Line	246°C
Engine Oil Pressure Gage	Red Line	65 PSI
	Green Arc	65 to 85 PSI
	Red Line	85 PSI
Carburetor Air Temp Indicator	Red Line	-30°C and + 54°C
	Green Arc	-30°C to -2°C
	Yellow Arc	-2°C to +32°C
	Green Arc	+32°C to +54°C

Courtesy Bell Helicopter Corporation

FIGURE 40.—Table of instrument markings.

1. Engine failure while hovering or on takeoff below 10 feet: A power failure is indicated by a sudden yawing of the ship to the left. In the event of such failure, do not reduce collective pitch. Apply right pedal to prevent excessive yawing. Apply collective pitch as necessary in order to cushion landing.

2. Engine failure during takeoff; altitude above 10 feet, below 500 feet: Caution: In order to effect a safe autorotation landing in the event of engine failure, takeoff operation should be conducted in accordance with the restrictions shown on Height–Velocity diagram. In the event of power failure during takeoff, the collective pitch must be initially lowered in order that the rotor speed may be maintained. The amount and duration of collective reduction depends upon the height above the ground at which the engine failure occurs. As the ground is approached, back cyclic and collective should be used as needed to decrease forward and vertical velocity. Ground contact should be established in a level attitude.

3. Engine failure above 500 feet altitude:
   a. Enter normal autorotation.
   b. Establish a steady glide of 61 to 63 mph IAS.
   c. At an altitude of about 50 feet, begin to steadily apply back cyclic stick to decrease forward speed.
   d. At approximately 10 feet, coordinate collective pitch with aft movement of cyclic stick to cushion landing. At ground contact, a level landing on the skid is accomplished.
   e. Avoid rapid lowering of collective pitch.
   f. In event of engine failure at night, do not turn on landing light above 1,000 feet above terrain in order to preserve battery power.

The emergency procedure just given is excerpted from the helicopter flight manual for one model of helicopter and should be used only for that particular model. Check the helicopter flight manual for each model of helicopter you fly to obtain the procedure to use in case of engine failure.

Performance information

This section should include information concerning (1) steady rates of climb and hovering ceilings together with the corresponding airspeeds and other pertinent information, including the calculated effect of altitude and temperature; (2) maximum wind allowable for safe operation near the ground; and (3) sufficient information to outline the limiting heights and corresponding speeds for safe landing after power failure.

The chart in figure 42 is an example taken from one manual from which rates of climb and best rate of climb

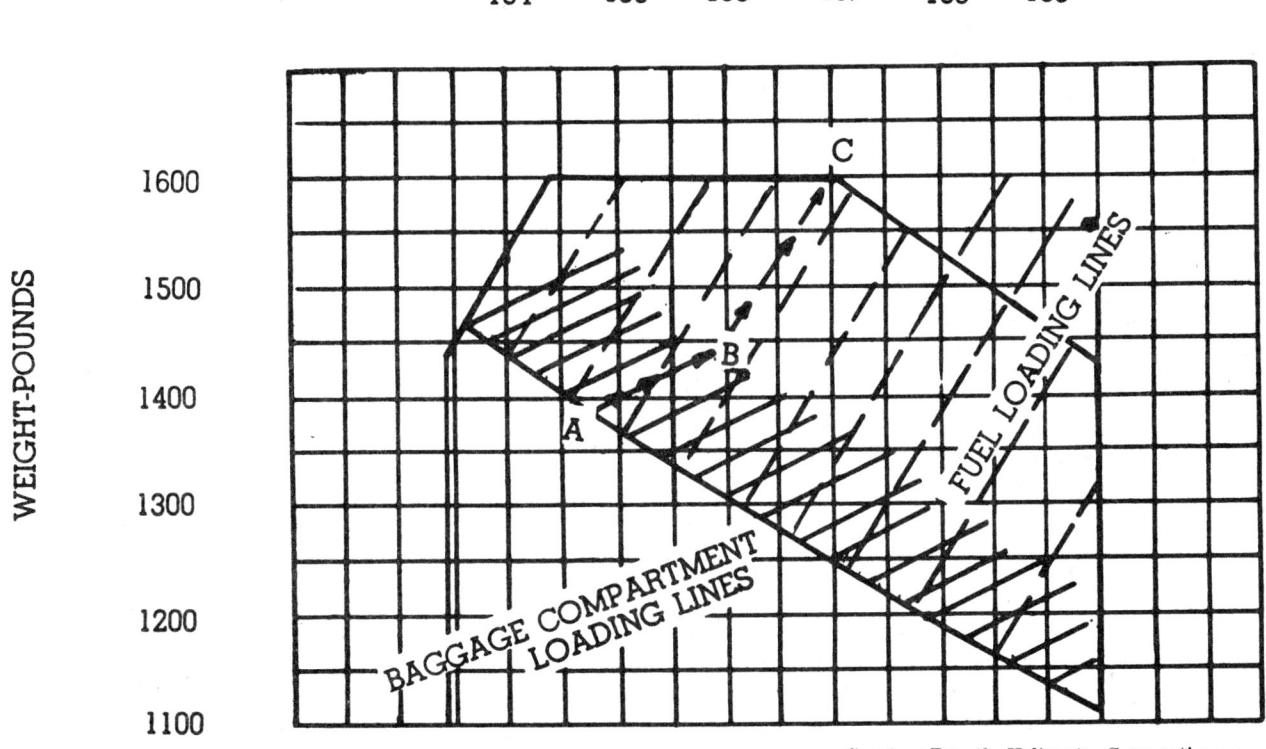

Courtesy Brantly Helicopter Corporation

FIGURE 41.—Loading chart.

airspeed can be computed. It can be used to compute the rate of climb for various density altitudes. The steps are as follows:

1. Compute the density altitude for the temperature/pressure altitude conditions by using a density altitude chart (fig. 51). Some flight manuals contain density altitude charts.

2. Locate the resulting density altitude along the left side of the chart (fig. 42), move horizontally from this point to the diagonal line, then vertically downward to the bottom line where the rate of climb in feet per minute is read. For example, if the computed density altitude is 1,000 feet, the rate of climb should be 1,250 feet per minute.

3. The best rate of climb airspeed can be found by moving horizontally from the density altitude point to the diagonal line in the chart on the right side, then moving vertically downward to the bottom line where the airspeed giving the best rate of climb in miles per hour can be read. In the example given in 2 above, it is approximately 57 MPH. At a density altitude of 8,500 feet, the best rate of climb airspeed is approximately 50 MPH and the rate of climb is approximately 425 feet per minute.

Maximum allowable wind for safe operations near the ground will be noted by a statement in most flight manuals similar to the following (excerpted from two helicopter flight manuals):

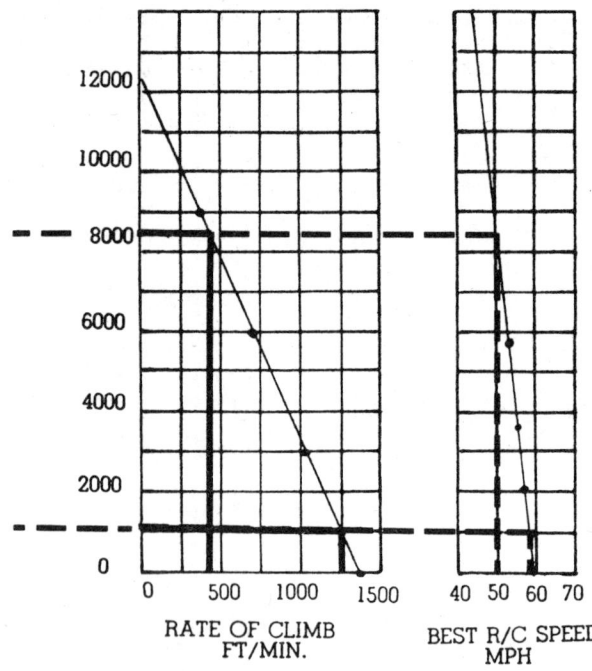

Courtesy Brantly Helicopter Corporation

FIGURE 42.—Rate of climb and best rate of climb speed chart.

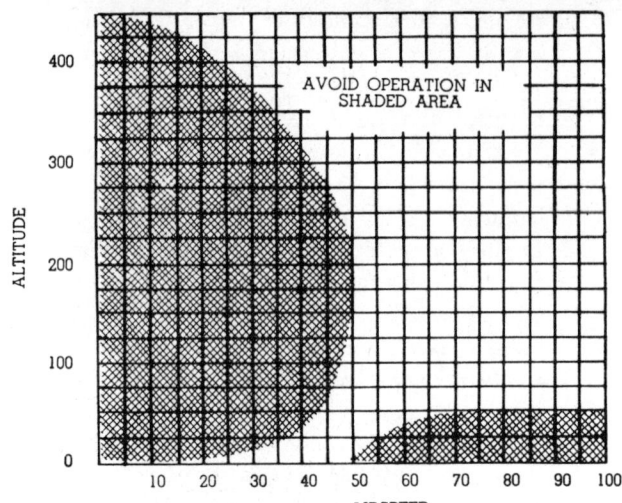

Courtesy Brantly Helicopter Corporation

FIGURE 43.—Airspeed vs. altitude limitations chart.

Rearward flight or hovering downwind can be conducted up to speeds of 23 mph IAS. Maximum possible operating wind velocities have not been established.

* * *

The maximum wind in which cross-wind takeoffs and landings and cross-wind and tail-into-the-wind hovering is safe is 20 mph; however, this is not to be considered the limiting value for flight as normal takeoffs, landings, and other maneuvers made close to the ground which do not require turning more than 45 degrees out of the wind may be made in much higher winds with the upper limit depending on experience and skill of the pilot.

Limiting heights and corresponding speeds for safe landing after power failure are generally incorporated in a chart called the Airspeed vs. Altitude Limitations Chart but often referred to as the Height–Velocity Curve or Diagram. This chart generally appears in the performance section of the helicopter flight manual, but occasionally may be found in the Operating Limitations section. Figures 43 and 44 represent such charts. These charts will be discussed in detail in a later chapter.

Placard information

All helicopters will generally have one or more placards displayed in conspicuous places that have a direct and important bearing on safe operation of the helicopter. These placards will generally appear in the helicopter flight manual in the operating limitations section under the heading of "Placards." Here are some examples from various flight manuals:

The following placard is installed on the baggage compartment door.

MAXIMUM BAGGAGE WEIGHT 50 POUNDS. SEE FLIGHT MANUAL FOR WEIGHT AND BALANCE DATA.

* * *

THIS HELICOPTER TO BE OPERATED IN COMPLIANCE WITH THE OPERATING LIMITATIONS SPECIFIED IN THE FAA-APPROVED ROTORCRAFT FLIGHT MANUAL.

* * *

SOLO FLIGHT PROHIBITED FROM LEFT SEAT.

A thorough understanding of the contents of the helicopter flight manual for the helicopter you fly will enable you to complete flights with maximum efficiency and safety.

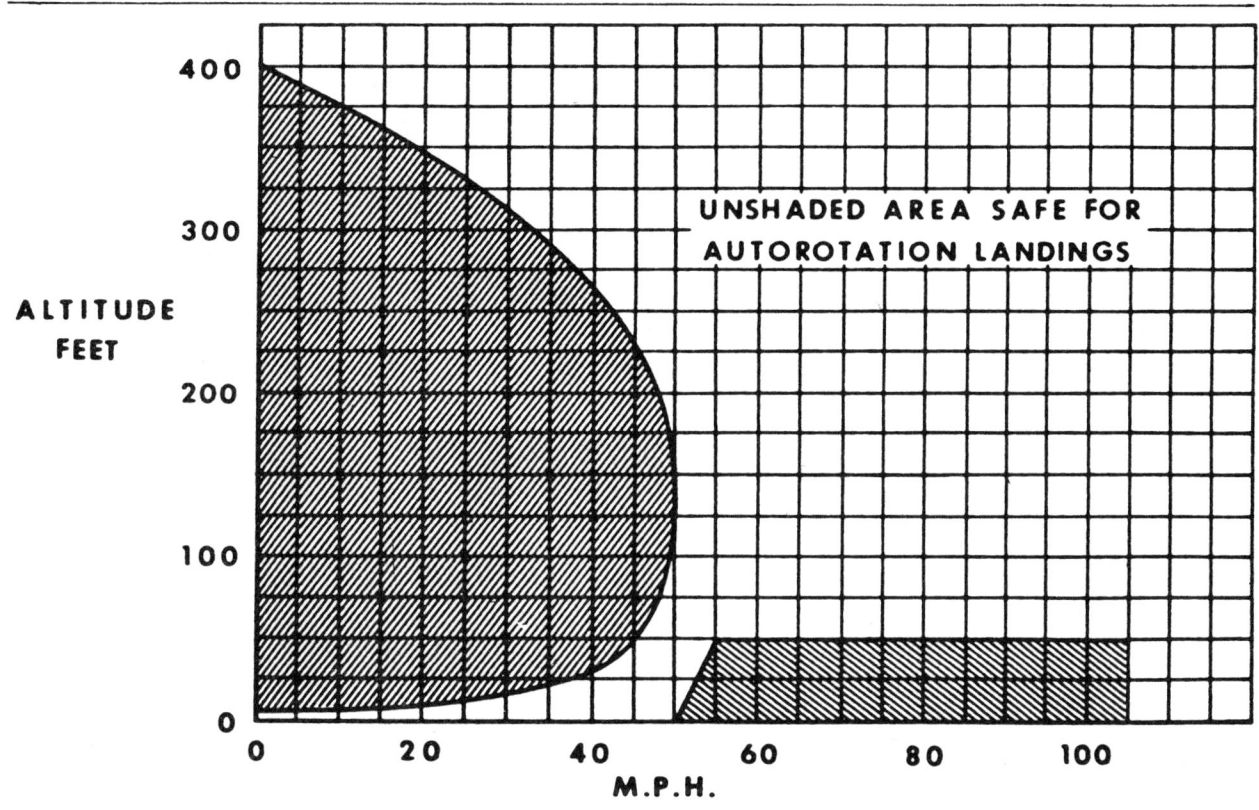

Courtesy Bell Helicopter Corporation

FIGURE 44.—Airspeed vs. altitude limitations chart.

Chapter 7. WEIGHT AND BALANCE

All helicopters are designed for certain limit loads and balance conditions. The pilot is responsible for making sure that the weight and balance limitations are met before takeoff. Any pilot who takes off in a helicopter that is not within the designed load and balance condition is not only violating the FAA regulations but is inviting disaster.

Three kinds of weight must be considered in the loading of every helicopter. These are empty weight, useful load, and gross weight.

Empty weight.—The weight of the helicopter including the structure, the powerplant, all fixed equipment, all fixed ballast, unusable fuel, undrainable oil, total quantity of engine coolant, and total quantity of hydraulic fluid.

Useful load (payload).—The weight of the pilot, passengers, baggage (including removable ballast), usable fuel, and drainable oil.

Gross weight.—The empty weight plus the useful load.

Maximum gross weight.—The maximum weight for which the helicopter is certificated for flight.

Some helicopter manufacturers use the term "basic weight" in determining the weight and balance of their helicopters. It includes the empty weight, as just defined, plus the weight of the drainable oil. Whenever the term "basic weight" is used in this handbook, it should be understood that this is the meaning. The term is not used in Federal Aviation Regulation, Part 27, Airworthiness Standards: Normal Category Rotorcraft.

Although a helicopter is certificated for a specified maximum gross weight, it will not be safe to take off with this load under all conditions. Conditions that affect take-off, climb, hovering, and landing performance may require the "off loading" of fuel, passengers, or baggage to some weight less than maximum allowable. Such conditions would include high altitudes, high temperatures, and high humidity, the combination of which determines the density altitude at any given place. Additional factors to consider are takeoff and landing surfaces, takeoff and landing distances available, and the presence of obstacles.

Because of the various adverse conditions that may exist, many times the helicopter pilot must decide the needs of the type mission to be flown and load his helicopter accordingly. For example, if all seats are occupied and maximum baggage is carried, gross weight limitations may require that less than maximum fuel be carried. On the other hand, if the pilot is interested in range, he may elect to carry a full fuel load but fewer passengers and less weight in baggage.

Balance

Not only must the pilot consider the gross weight of his helicopter, he must also determine that the load is arranged to fall within the allowable center-of-gravity range specified in the helicopter weight and balance limitations contained in the helicopter flight manual. The center of gravity, often referred to as the "CG," is the point where the helicopter is in balance—the point at which all the weight of the system is considered to be concentrated. If the helicopter were suspended by a string attached to the center-of-gravity point, the helicopter fuselage would remain parallel to the ground much as a perfectly balanced teeter-totter (seesaw). The allowable range in which the CG may fall is referred to as the CG range. The exact location and length of this range is specified for each helicopter, but it usually extends a short distance fore and aft of the main rotor mast. For most types of helicopters, the location of the CG must be kept within much narrower limits than in airplanes—less than 3 inches in some cases.

The ideal condition is to have the helicopter in such perfect balance that the fuselage will remain horizontal in hovering flight, with no cyclic pitch control necessary except that which may be made necessary by wind. The fuselage acts as a pendulum suspended from the rotor. Any change in the center of gravity changes the angle at which it hangs from this point of support. If the center of gravity is located directly under the rotor mast, the helicopter hangs horizontal; if the center of gravity is too far aft of the mast, the helicopter hangs with nose tilted up; and if the center of gravity is too far forward of the mast, the nose tilts down (fig. 45). *Out of balance loading of the helicopter makes control more difficult and decreases maneuverability* since cyclic stick travel is restricted in the direction opposite to CG location. Because helicopters are relatively narrow and high sideward speeds will not be attained, lateral balance presents no problems in normal flight instruction and passenger flights, although some light helicopters specify the seat from which solo flight will be made. However, if external loads are carried in such a position that a large lateral displacement of the cyclic stick is required to maintain level flight, fore and aft cyclic stick movement will be very limited.

CG forward of allowable limits

This condition may arise in a two-place helicopter when a heavy pilot and passenger take off without baggage or proper ballast located aft of the rotor mast. The condition will become worse as the flight progresses due to fuel consumption, if the main fuel tank is also located aft of the rotor mast.

The pilot will recognize this condition after coming to a hover following a vertical takeoff. The helicopter will have a nose-low attitude and an excessive rearward displacement of the cyclic stick will be required to maintain a hover in a no-wind condition if hovering flight can be maintained at all (fig. 45). Flight under this condition *should not be continued* since the possibility of running out of rearward cyclic control will increase rapidly as fuel is consumed, and the pilot may find it impossible to decelerate sufficiently to bring the helicopter to a stop. Also, in case of engine failure and the resulting autorotation, sufficient cyclic control may not be available to flare properly for the landing.

Hovering into a strong wind will make this condition less easily recognizable since less rearward displacement of the cyclic control will be required than when hovering in a no-wind condition. Therefore, in determining if a critical balance condition exists, the pilot should consider the wind velocity in which he is hovering and its relation to the rearward displacement of the cyclic stick.

CG aft of maximum limits

Without proper ballast in the cockpit, this condition may arise when (1) a lightweight pilot takes off solo with a full load of fuel located aft of the rotor mast; (2) a lightweight pilot takes off with maximum baggage allowed in a baggage compartment located aft of the rotor mast; or (3) a lightweight pilot takes off with a combination of baggage and substantial fuel where both are aft of the rotor mast.

The pilot will recognize this condition after coming to a hover following a vertical takeoff. The helicopter will have a tail-low attitude and an excessive forward displacement of the cyclic stick will be required to maintain a hover in a no-wind condition if a hover can be maintained at all. If there is a wind, an even greater forward displacement will be required.

If flight is continued in this condition, the pilot may find it impossible to fly in the upper allowable airspeed range due to insufficient forward cyclic displacement to maintain a nose-low attitude. This particular condition may become quite dangerous if gusty or rough air accelerates the helicopter to a higher airspeed than forward cyclic control will allow. The nose will start to rise and full forward cyclic stick may not be sufficient to hold it down or to lower it once it rises.

Weight and balance information

When a helicopter is delivered from the factory, the empty weight, empty weight CG, and useful load for each particular helicopter are noted on a weight and balance data sheet included in the helicopter flight manual. These quantities will vary for different helicopters of a given series depending upon changes or variations in the fixed equipment included in each helicopter when delivered.

If, after delivery, additional fixed equipment is added, or if some is removed, or a major repair or alteration is made which may affect the empty weight, empty weight

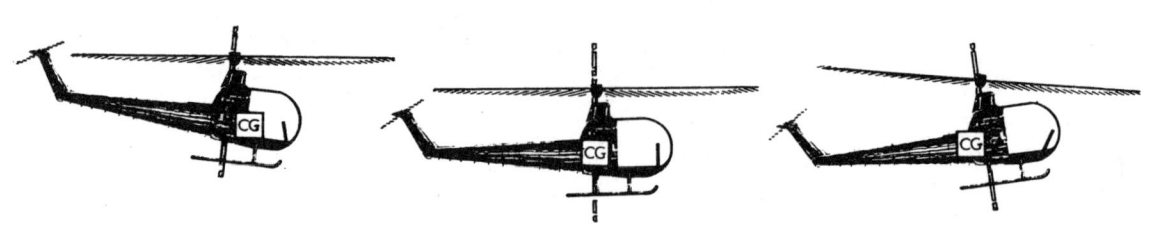

Figure 45.—Effect of center of gravity on cyclic stick position and helicopter attitude during hovering flight.

CG, or useful load, the weight and balance data must be revised to reflect these new values. All weight and balance changes will be entered in the appropriate aircraft record. This generally will be the aircraft logbook. The *latest* weight and balance data should be used in computing all loading problems.

Sample weight and balance problems

In loading a helicopter for flight the problem is twofold:

- Is the gross weight within the maximum allowable gross weight?
- Does the helicopter meet balance requirements, i.e., is the CG within the allowable CG range?

To answer the first question, merely add the weight of the items comprising the useful load (pilot, passengers, fuel, oil, and baggage) to the empty weight of the helicopter. (Obtain the latest empty weight information from the appropriate aircraft record.) Then check the total weight obtained to see that it does not exceed maximum allowable gross weight. If basic weight is used in computing weight and balance, then the weight of the oil is included with this weight.

To answer the second question, use the loading chart or loading table in the helicopter flight manual for the particular helicopter that is being flown.

Sample problem 1.—Determine if the gross weight and center of gravity are within allowable limits under the following loading conditions for a helicopter based on the loading chart in figure 46.

	Pounds
Basic weight	1,040
Seat load:	
Pilot	135
Passenger	200
Baggage	25
Fuel (30 gallons)	180

SOLUTION: To use the loading chart for the helicopter in this example, the items comprising the useful load must be added to the basic weight in a certain order. The maximum allowable gross weight is 1,600 pounds.

	Pounds
Basic helicopter weight	1,040
Seat load (pilot—135 pounds; passenger—200 pounds)	335
Subtotal	1,375 (point A)
Baggage compartment load	25
Subtotal	1,400 (point B)
Fuel load	180
Total weight	1,580 (point C)

The total weight of the helicopter does not exceed the maximum allowable gross weight. By following the sequence of steps (locating points A, B, and C) on the loading chart, it is found that the CG is within allowable limits (fig. 46).

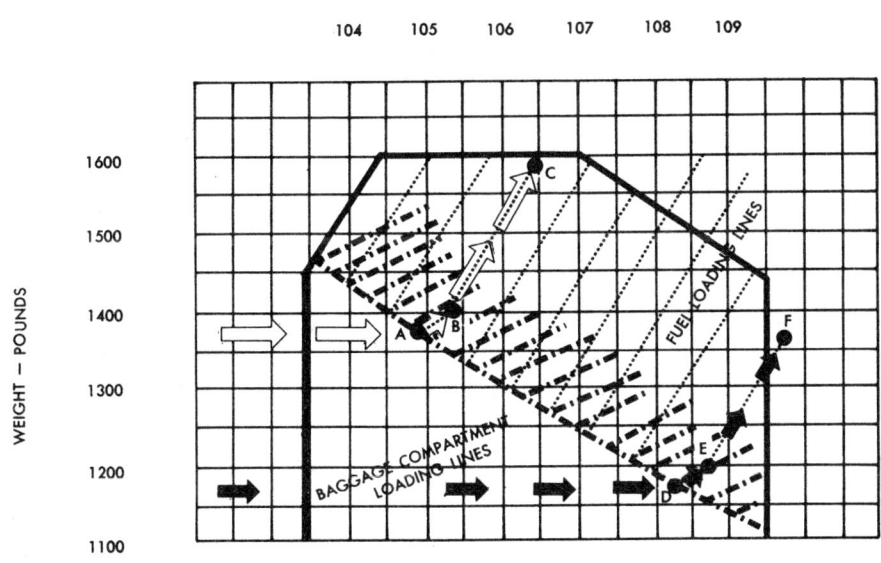

FIGURE 46.—Loading chart illustrating the solution to sample problems 1 and 2.

Sample problem 2.—For this example, assume that the pilot in sample problem 1 discharges his passenger after using only 20 pounds of fuel.

SOLUTION: *Pounds*

Basic helicopter weight.................... 1,040
Seat load (pilot—135 pounds)...........   135
                                           _____
    Subtotal........................ 1,175 (point D)
Baggage compartment load.................   25
                                           _____
    Subtotal........................ 1,200 (point E)
Fuel load...............................   160
                                           _____
    Total weight.................... 1,360 (point F)

Although the total weight of the helicopter is well below the maximum allowable gross weight, the CG falls outside of the aft allowable limit as defined by the loading chart (fig. 46).

This example illustrates the importance of reevaluating the balance problem in a helicopter whenever a change is made in the loading. In most airplanes, the discharging of a passenger would have little effect on the CG. Because of this fact, the airplane pilot may not be aware of the critical loading problems that he must constantly keep in mind as a helicopter pilot. If the pilot in this example takes off again after discharging his passenger, he will find that he has insufficient forward cyclic stick displacement to hover in a strong wind; it may be impossible to fly in the upper airspeed range due to insufficient forward cyclic displacement to maintain a nose-low atti-tude; and a dangerous situation will exist if gusty or rough air accelerates the helicopter to a higher airspeed than forward cyclic control will allow.

Just the opposite situation from that illustrated by sample problems 1 and 2 could exist if a very heavy pilot proceeded solo on a flight until a substantial amount of fuel was used then stopped to pick up a very heavy passenger and proceeded on his flight without refueling. During the solo portion of the flight, the helicopter CG would be within allowable limits. During the second portion of the flight with the passenger, the CG would be forward of allowable limits. If this pilot did take off with the passenger in this example, he would find that he had insufficient rearward cyclic stick displacement to slow the aircraft to a hover and, in case of an autorotative landing, insufficient back cyclic stick available to flare the helicopter for a landing.

Determining the CG without the use of a loading chart or table

An alternate method of computing the center of gravity is to use the arm-weight-moment computation method. To use this method, we must know the empty weight and empty weight CG (obtained from the appropriate aircraft record) or the basic weight and basic weight CG, and the weight and distance from the datum line of each portion of the useful load—pilot and passengers, usable fuel, baggage, and drainable oil (unless basic weight is used).

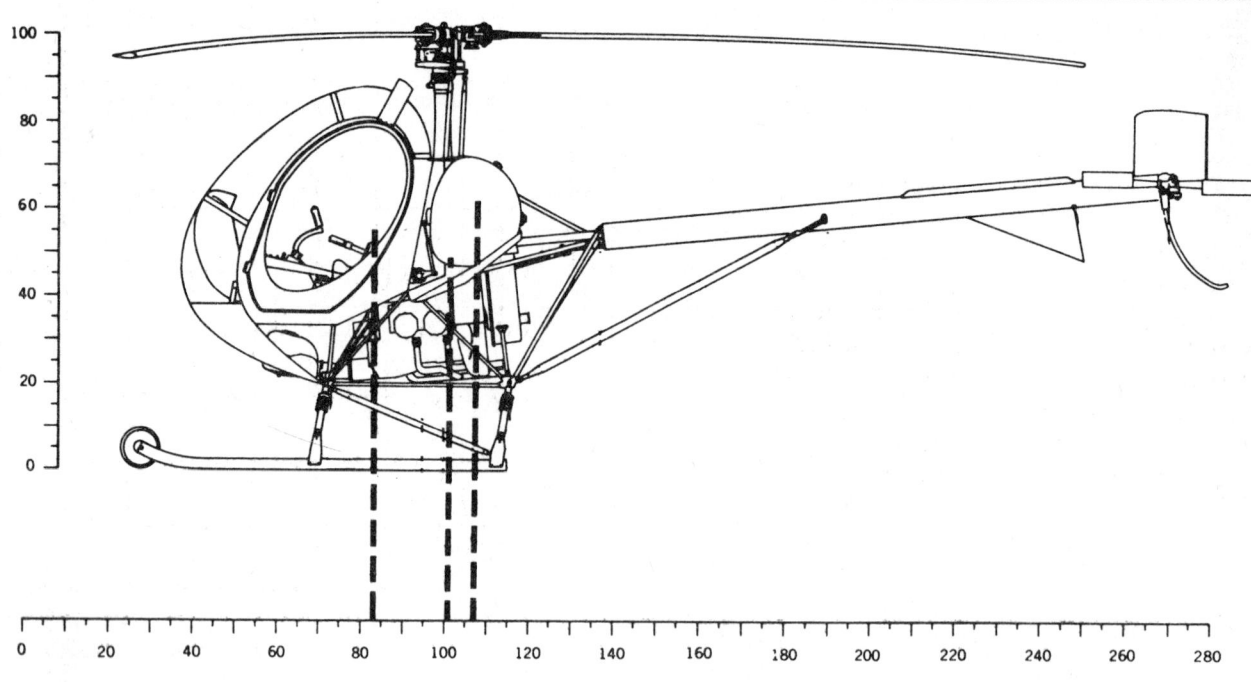

Courtesy Hughes Tool Company, Aircraft Division

FIGURE 47.—Datum line forward of helicopter.

The following formulas will be used:

$$\text{Weight} \times \text{arm} = \text{Moment}$$
$$\text{Moment} + \text{moment} = \text{Total Moment}$$
$$\frac{\text{Total moment}}{\text{Total weight}} = \text{CG}$$

Some manufacturers choose the datum line at or ahead of the most forward structural point on the helicopter in which case all moments are positive. Other manufacturers choose the datum line at some point in the middle of the helicopter in which case moments produced by weight ahead of the datum line are negative and moments produced by weight aft of the datum line are positive. A sample problem will be shown for each type.

If external loads are carried, this is the method by which the center of gravity will normally be computed.

Sample Problem 3.—For this example, the datum line is chosen ahead of the most forward structural point on the helicopter (fig. 47).

Item	Weight (pounds)	Arm (inches)	Moment (inch-pounds)
Basic weight...............	1,004	101.0	101,404
Fuel (25 gallons)..........	150	107.0	16,050
Seat load.................	330	83.9	27,687
TOTALS..............	1,484		145,141

$$CG = \frac{\text{Total moment}}{\text{Total weight}} = \frac{145{,}141}{1{,}484} = 97.8 \text{ inches aft of datum line}$$

Since the approved center-of-gravity limits are station 95 (95 inches from datum line) and station 100 and the maximum allowable gross weight is 1,600 pounds, the helicopter meets the weight and balance requirements for flight.

Figures 48 and 49 show how the charts in the helicopter flight manual would normally be used to compute the CG in this helicopter. To compute the CG from these charts, the first step is to determine the total weight of the helicopter (in pounds) and the total moment (in thousands of inch-pounds).

1. The empty weight and empty weight CG or basic weight and basic weight CG are obtained from the latest weight and balance information in the helicopter flight manual. In this case, 1,004 pounds and 101.4.

2. The total moments for the fuel load and seat load are found from the loading chart (fig. 48). Locate the diagonal line indicating fuel quantities and stations. Draw a horizontal line from the 25-gallon mark to the left side of the chart where the moment is read to be 16.

3. Locate the point representing the seat load along the bottom of the chart. From this point, draw a vertical line until it intersects the diagonal line marked "Pilot and Passenger." From this point of intersection, draw a horizontal line to the left side of the chart where the moment

is read as 27.7. Tabulating these results and totaling, we have the following:

Item	Weight (pounds)	Moment (thousands of inch-pounds)
Basic weight...............	1,004	101.4
Fuel (25 gallons)..........	150	16.0
Seat load.................	330	27.7
TOTALS................	1,484	145.1

The second step is to locate the point on the center-of-gravity chart (fig. 49) represented by the total weight and total moment just computed. This point falls within the center-of-gravity envelope; therefore, the helicopter is loaded within center-of-gravity and weight limits.

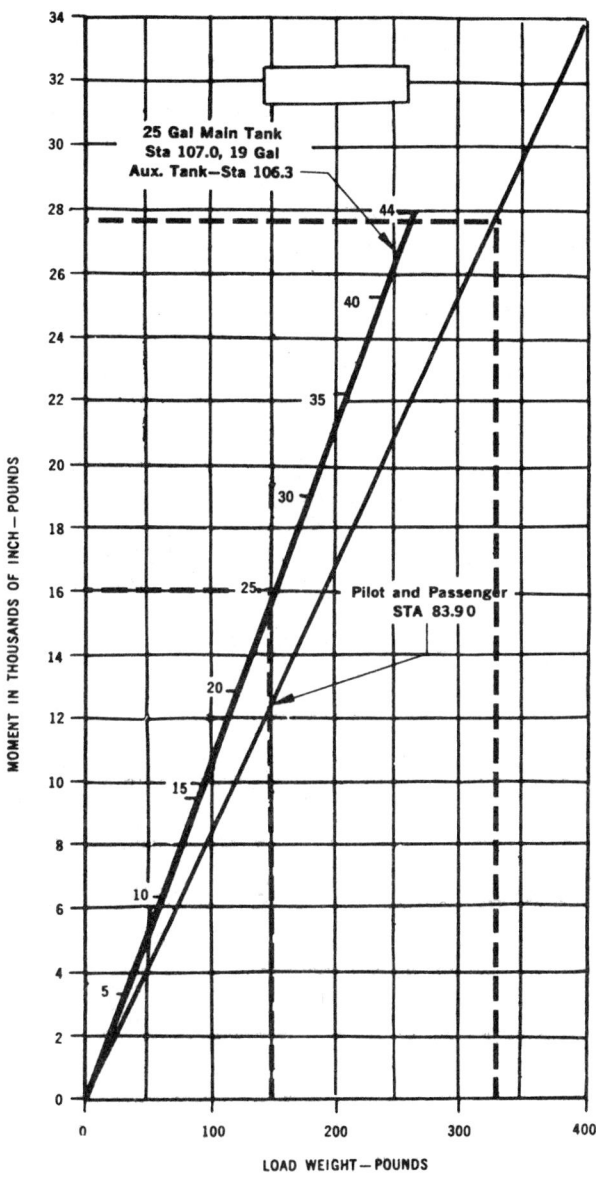

FIGURE 48.—Loading chart.

Sample Problem 4.—In this example, the datum line is chosen at a point in the middle of the helicopter (fig. 50).

Item	Weight (pounds)	Arm (inches)	Moment (inch-pound) Positive	Negative
Empty weight.....	1,820	+6	10,920	....
Fuel (41 gallons)...	246	+2	492	....
Seat load.........	330	−31	....	10,230
TOTALS.......	2,396		11,412	10,230

Total moment = 11,412 − 10,230 = 1,182

$$CG = \frac{\text{Total moment}}{\text{Total weight}} = \frac{1,182}{2,396} = +0.5 \text{ inches aft of datum line}$$

Since the approved center-of-gravity limits are 3 inches forward of station 0 to 4 inches aft of station 0, and the maximum allowable gross weight is 2,850 pounds, the helicopter meets the weight and balance requirements for flight.

Because the total positive moment exceeds the total negative moment, the center of gravity is aft of the datum line. Had the negative moment exceeded the positive moment, the CG would be forward of the datum line.

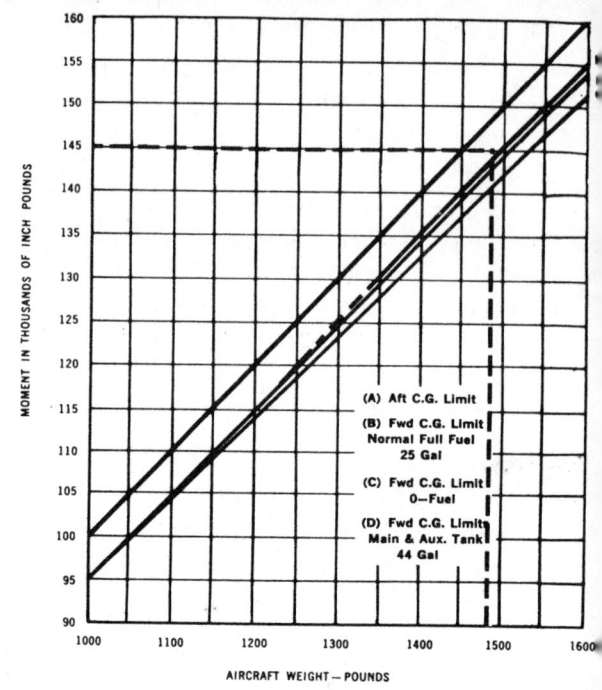

FIGURE 49.—Center-of-gravity chart.

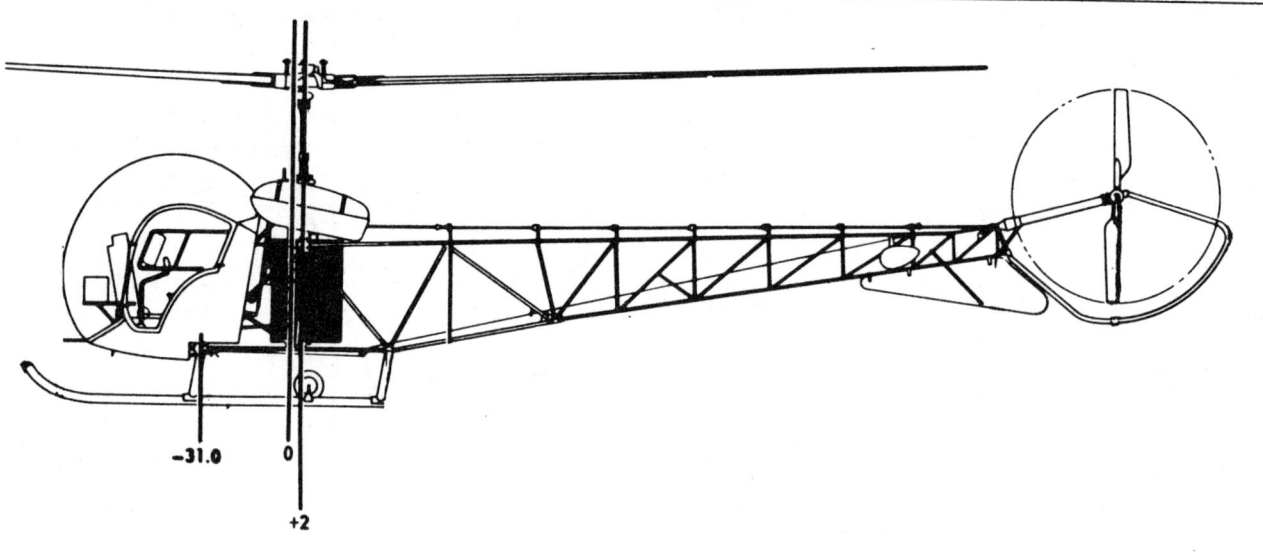

FIGURE 50.—Datum line near rotor mast.

Chapter 8. HELICOPTER PERFORMANCE

Assuming that a helicopter engine and all components are operating satisfactorily, the performance of the helicopter is dependent on three major factors:

- Density altitude (air density)
- Gross weight
- Wind velocity during takeoff, hovering, and landing

Air density

Air, like liquids and other gases, is a fluid. Because it is a fluid, it flows and changes shape under pressure. Air is said to be "thin" at high altitudes; that is, there are fewer molecules per cubic foot of air at 10,000 feet than at sea level. The air at sea level is "thin" when compared to air compressed to 30 pounds of pressure in an automobile tire. A cubic inch of air compressed in an automobile tire is *denser* than a cubic inch of "free" air at sea level.

For example, in a stack of blankets, the bottom blanket is under pressure of all blankets above it. As a result of this pressure, the bottom blanket may be squeezed down until it is only one-tenth as bulky as the fluffy blanket on top. There is still just as much wool in the bottom blanket as there is in the one on top, but the wool in the bottom blanket is 10 times more dense. If the second blanket from the bottom of the stack were removed, a force of 15 pounds might be required to pull it out. The second blanket from the top may require only 1 pound of force. In the same way, air layers near the earth's surface have much greater density than air layers at higher altitudes. The lower the elevation of the earth's surface, the greater the density of the air layers. For example, the layer of air at sea level would be denser than the layer of air at the earth's surface at Denver, Colo., at approximately 1 mile above sea level.

The above principle may be applied in flying aircraft. At lower levels, the rotor blade is cutting through *more* and *denser* air, which offers more support (lift) and increases air resistance. The same amount of power, applied at higher altitudes where the air is *thinner* and *less dense*, propels the helicopter faster.

Density altitude

Density altitude refers to a theoretical air density which exists under standard conditions of a given altitude.

Standard conditions at sea level are:

Atmospheric pressure—29.92 in. of Hg (inches of mercury)

Temperature—59° F. (15° C.)

Standard conditions at any higher altitude are based on:

Atmospheric pressure (reduced to sea level)—29.92 in. of Hg

Temperature—59° F. (15° C.) minus 3½° F. (2° C.) per 1,000 feet elevation

For example, if the atmospheric pressure (reduced to sea level) at an airport located 5,000 feet above sea level is 29.92 inches of mercury and the temperature is 59° — $(3.5° \times 5) = 41.5°$ F. (5° C.), the air density is standard at that altitude. (The actual barometric pressure at an elevation of 5,000 feet under these conditions would be approximately 24.92 inches of mercury since atmospheric pressure decreases approximately 1 inch per 1,000-foot increase in altitude. The average temperature decrease per 1,000-foot increase in altitude is 3.5° F.).

Figure 51 shows a density altitude chart. If we locate the 5° vertical line along the bottom of the chart, follow this line up to its intersection with the 5,000-foot diagonal line, then follow the horizontal line to the left side of the chart, we read a density altitude of 5,000 feet.

The four factors which affect density altitude are altitude, atmospheric pressure, temperature, and moisture content of the air.

Altitude

We have already seen the effects of altitude on air density in the first section of this chapter. The greater the elevation of an airport or landing area, the less the atmospheric pressure and, consequently, the less dense the air. The less dense the air, the greater the density altitude. What is the result when operating at a high density altitude? Helicopter performance is decreased (fig. 52). It can be seen from the density altitude chart that, as altitude increases, density altitude increases.

Atmospheric pressure

The atmospheric pressure at an airport or landing area at a given elevation can change from day to day—sometimes a very noticeable amount which, when combined with other factors, could be significant. The lower the pressure at a given elevation, the less dense the air; the

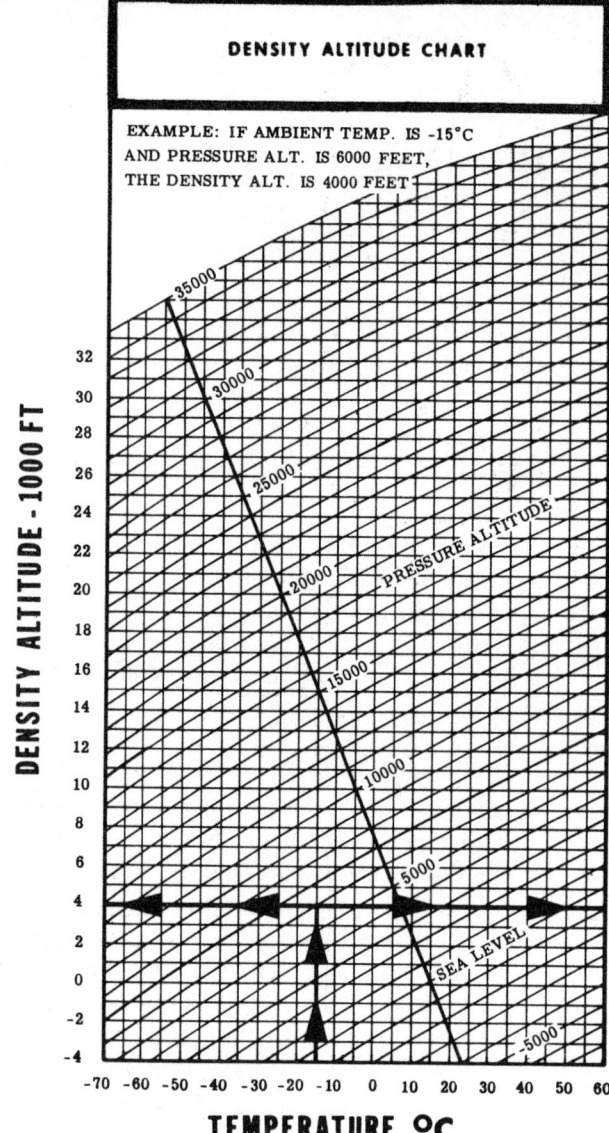

FIGURE 51.—Density altitude chart.

Temperature

Even when elevation and pressure remain constant, great changes in air density will be caused by temperature changes. The same amount of air that occupies 1 cubic inch at a low temperature will expand and occupy 2, 3, or 4 cubic inches as the temperature rises. Therefore, as temperature increases, air becomes less dense, density altitude is increased, and the helicopter performance decreases (fig. 53). A study of figure 51 easily reveals that, as temperature increases, density altitude increases since the pressure altitude lines slope upward to the right.

We have already used the density altitude chart to find the density altitude at an elevation of 5,000 feet under standard atmospheric conditions for that elevation, that is, atmospheric pressure (reduced to sea level) 29.92 inches of mercury, temperature 41.5° F. (5° C.).

When the atmospheric pressure, reduced to sea level, at a given elevation is 29.92, the pressure altitude is the same as the given elevation.

What would be the density altitude at this same elevation if the pressure altitude is still 5,000 feet but the temperature is 95° F. (35° C.)? Locate the 35° vertical line at the bottom of the chart, follow this line up to its intersection with the 5,000-foot pressure altitude (diagonal) line, then move horizontally to the left side of the chart where a density altitude of 8,400 feet is read. A helicopter operating at this elevation under these conditions would be flying in air with a density equivalent to that at the 8,400-foot level. Therefore, the performance of the helicopter would be as though it were flying at the 8,400-foot level rather than the 5,000-foot level.

Moisture

When temperature and pressure are constant, changes in the moisture content of the air will change air density. Water vapor weighs less than dry air. Therefore, as the moisture content of the air increases, air becomes less dense; density altitude is increased with a resultant decrease in helicopter performance (fig. 54.) *The higher the temperature the greater the amount of moisture the air can hold.* Relative humidity, which is expressed as a percent, is the ratio of the amount of moisture in the air to the amount it is capable of absorbing at a given temperature. The moisture content of the air at a relative humidity of 80 percent and a temperature of 100° F. will be much greater than with a relative humidity of 80 percent and a temperature of 50° F. The greatest decrease in air density (increase in density altitude) due to moisture content will be at a high temperature.

The density altitude chart (fig. 51) does not take the moisture content of the air into consideration. *It should*

less dense the air, the higher the density altitude and, as a result, the less performance the helicopter will have.

The daily and seasonal variations in atmospheric pressure at a given place will not have as significant effect on the density altitude as the daily and seasonal variations of temperature and moisture.

The density altitude chart is based on pressure altitude, not indicated altitude (fig. 51). To determine the pressure altitude at any given place, if an altimeter is available, adjust the altimeter setting to 29.92 and read the pressure altitude directly from the altimeter. However, do not forget to reset the altimeter to the current altimeter setting if available, or to field elevation if an altimeter setting is not available.

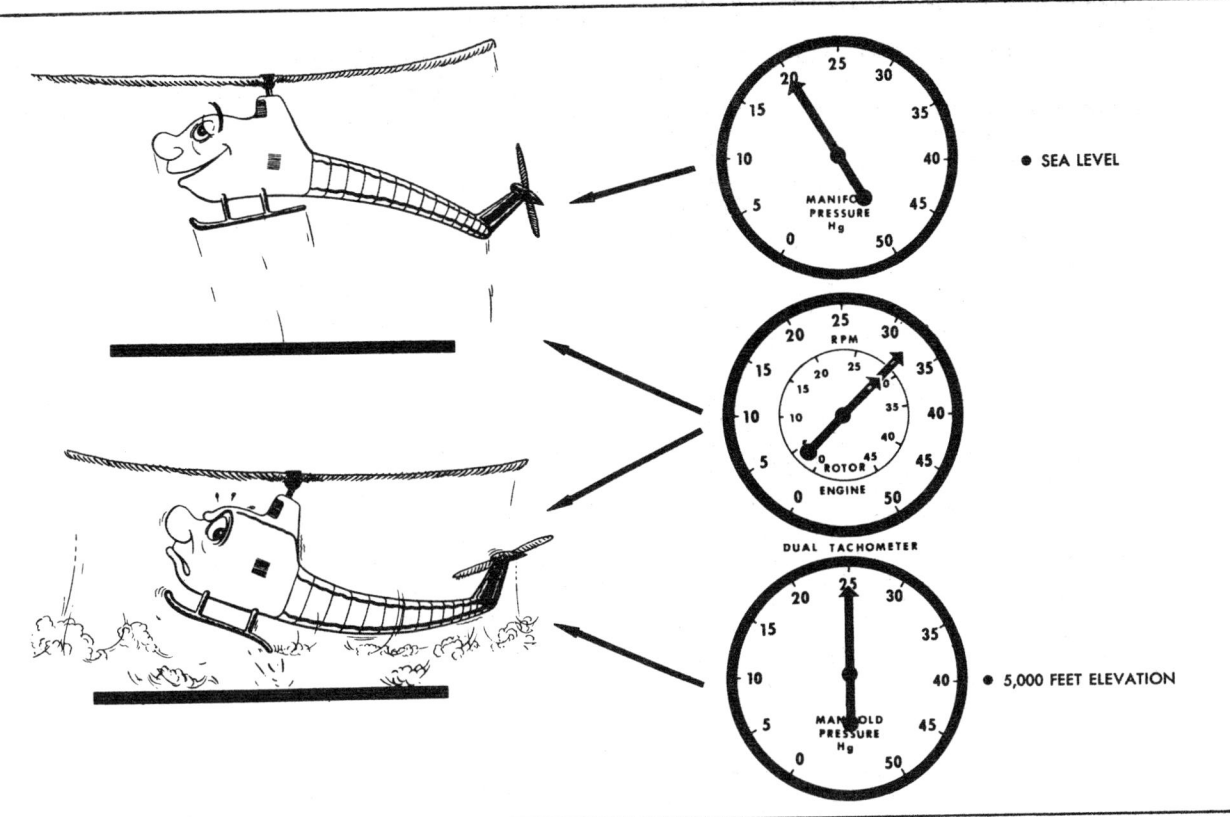

FIGURE 52.—Helicopter performance is reduced at high elevations.

be remembered that the actual density altitude can be much higher than that computed from this chart if the air contains a high moisture content. The importance of this added effect of moisture on helicopter hovering performance will be seen shortly.

High density and low density altitude conditions

The terms high density altitude and low density altitude should be thoroughly understood. In general, high density altitude refers to thin air; low density altitude refers to dense air. Therefore, those conditions that result in thin air—high elevations, high temperatures, high-moisture content, or some combination thereof—would be referred to as high density altitude conditions; those conditions that result in dense air—low elevations, low temperatures, low moisture content, or some combination thereof—would be referred to as low density altitude conditions. It is important to note that high density altitudes may be present at low elevations on hot days with high moisture content in the air.

EFFECT OF HIGH DENSITY ALTITUDES ON HELICOPTER PERFORMANCE

High elevations, high temperatures, and high-moisture content, all of which contribute to a high density altitude condition, lessen helicopter performance. Because the difference between the power available and the power required is so small for a helicopter, particularly in hovering flight, density altitude is of even greater importance to the helicopter pilot than it is to the airplane pilot. Helicopter performance is reduced because the thinner air at high density altitudes reduces the amount of lift of the rotor blades. Also, the (unsupercharged) engine does not develop as much power because of the thinner air and the decreased atmospheric pressure.

Hovering flight

High density altitudes reduce the hovering capabilities of the helicopter (fig. 55). Under any given load condition, the higher the density altitude, the lower the hovering ceiling; that is, the elevation at which the helicopter will be able to hover will be lowered as the density altitude increases.

Figure 56 gives the hovering ceiling in ground effect for one helicopter at various gross weights and temperatures both in dry air and air at 80 percent relative humidity. The following previously established points should be easily recognized from this chart:

1. *An increase in temperature decreases the hovering ceiling.*—For example, at 1,600 pounds gross weight in

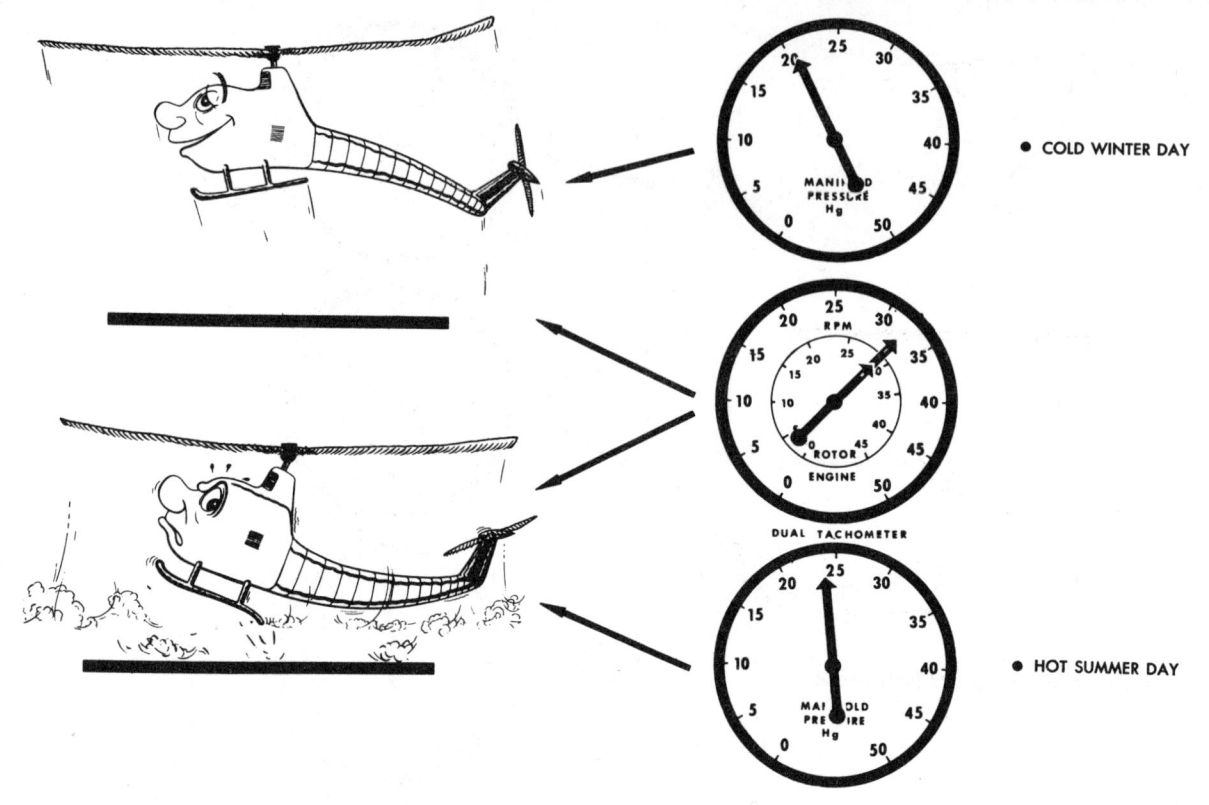

FIGURE 53.—High temperature reduces helicopter performance.

dry air, as the temperatures increases from −20° F. to 100° F., the hovering ceiling decreases from 6,700 feet to 3,000 feet. At 1,600 pounds in 80 percent relative humidity, as the temperature increases from −20° F. to 100° F., the hovering ceiling decreases from 6,500 feet to 1,300 feet.

2. *As the amount of moisture in the air increases, the hovering ceiling decreases.*—For example, at 1,600 pounds gross weight and 100° F., as the moisture content changes from dry air to 80 percent relative humidity, the hovering ceiling decreases from 3,000 feet to 1,300 feet; at 1,600 pounds and 20° F., as the moisture changes from dry air to 80 percent relative humidity, the hovering ceiling decreases from 5,500 feet to only 5,200 feet.

3. *The higher the temperature, the greater the amount of moisture which the air can hold.*—At 1,600 pounds gross weight, and temperatures of −20° F., 20° F., 60° F., and 100° F., the change in hovering ceiling from dry air to 80 percent relative humidity is 200 feet, 300 feet, 400 feet, and 1,700 feet, respectively. This indicates that the amount of moisture in the air at 100° F. and 80 percent relative humidity is much greater than the amount present at 60° F. and below.

Figure 57 illustrates the hovering ceiling in ground effect for temperature and gross weight variations for one helicopter but does not reflect the effect that moisture content has on the performance. For example, at a gross weight of 1,550 pounds and a temperature of 100° F., the hovering ceiling is a pressure altitude of 4,100 feet.

Takeoff

For any given gross weight, the higher the density altitude at point of departure, the more power that is required to make a vertical takeoff to a hover (fig. 55). In fact, under certain gross weight and density altitude conditions, a helicopter may not have sufficient power to lift off vertically in which case, if takeoff is made, it would have to be a running takeoff.

Figure 58 shows a chart that is used to compute the takeoff distance required to clear a 50-foot obstacle under various gross weight, pressure altitude, and temperature conditions. A brief study of the chart immediately reveals the previously established points, that is, as gross weight, altitude, and temperature increase, the takeoff performance decreases. This chart is used in the following way:

1. In the first column, locate the helicopter gross weight.

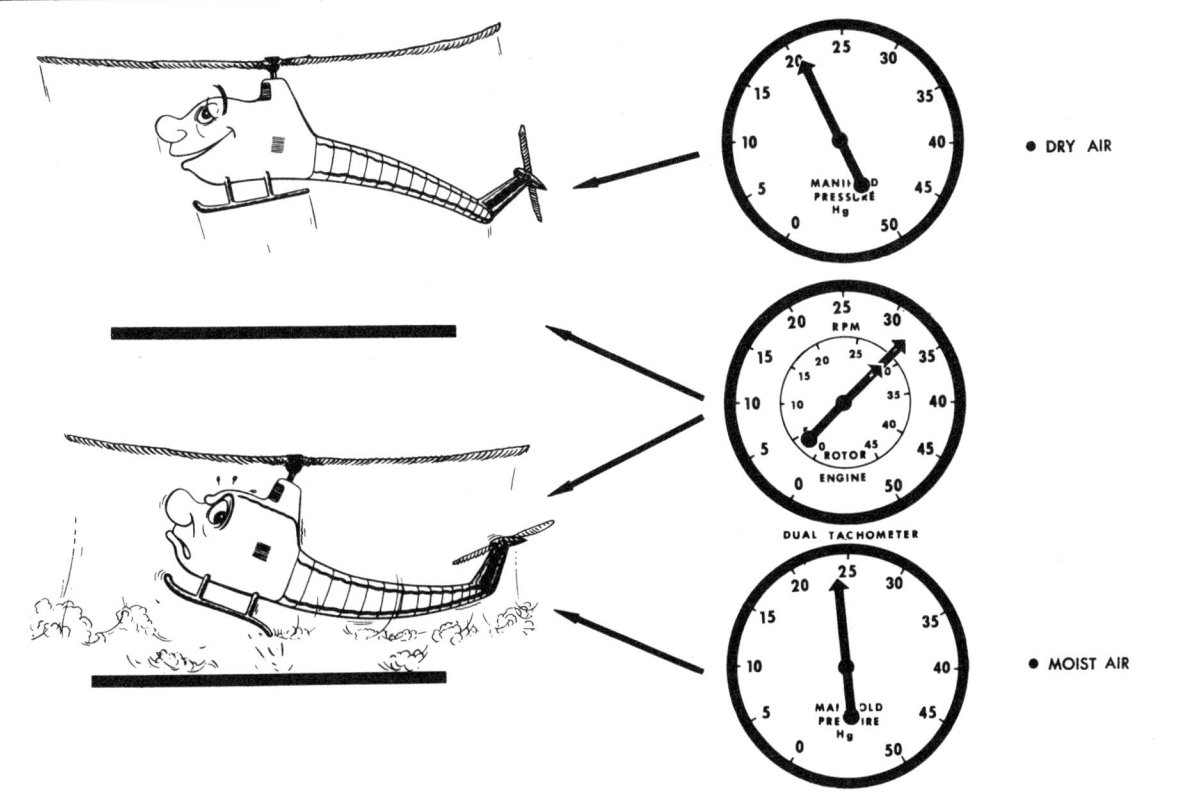

FIGURE 54.—High humidity (especially on hot days) reduces helicopter performance.

2. In the second column, opposite the gross weight, locate the pressure altitude at point of takeoff.

3. Follow this pressure altitude row out to the column headed by the temperature at point of takeoff.

4. The figure at the intersection of the pressure altitude row and the temperature column is the number of feet required for this particular helicopter to take off and clear a 50-foot obstacle.

5. If the gross weight, pressure altitude, or temperature, or any combination of the three, fall between the listed values, the interpolation process will have to be used to compute the distance.

Sample Problem 1.—What distance is required to clear a 50-foot obstacle if the helicopter gross weight is 2,500 pounds, the pressure altitude is 6,000 feet and the temperature is 59° F.?

SOLUTION:

1. In the first column, locate 2,500.

2. In the second column opposite 2,500, locate 6,000.

3. Follow this 6,000 row out to the column headed by 59° F. where you read 848.

4. The distance required to clear a 50-foot obstacle under these conditions then is 848 feet.

Sample Problem 2.—What distance is required to clear a 50-foot obstacle if the helicopter gross weight is 2,850 pounds, the pressure altitude is 5,000 feet, and the temperature is 95° F.? The solution requires interpolation.

SOLUTION:

1. In the first column, locate 2,850.

2. In the second column opposite 2,850, locate the 4,000- and 6,000-foot rows.

3. Follow each of these rows out to the column headed by 95° F., where you read 1,102 and 1,538, respectively.

4. Since 5,000 feet falls midway between 4,000 and 6,000, we assume that the takeoff distance at this altitude falls midway between 1,102 and 1,538.

5. By taking half of the difference of the two distances and adding it to 1,102 (or subtracting it from 1,538), the distance required to clear a 50-foot obstacle under the conditions of the problem is 1,320 feet.

Since this chart does not take into consideration the decrease in air density due to moisture content, *takeoff distances may be even greater than those computed from the table.*

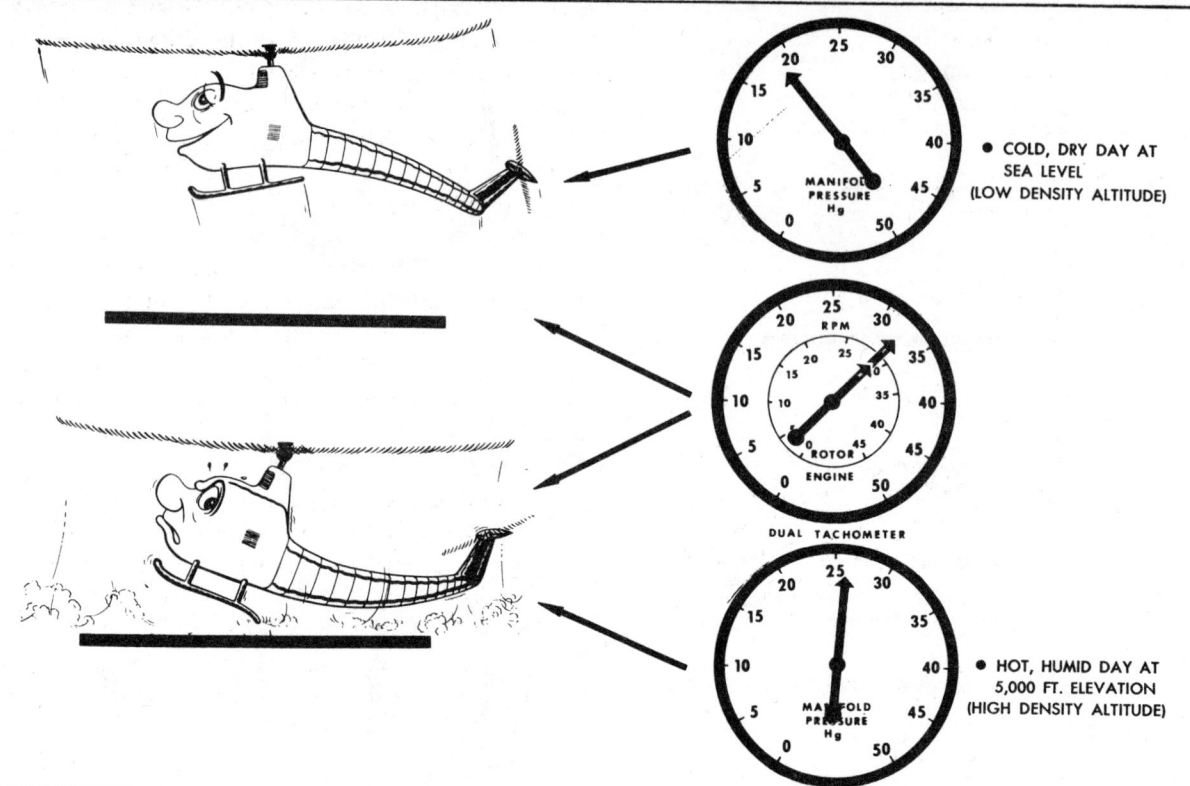

FIGURE 55.—High density altitudes reduce helicopter performance.

Gross Weight Lbs.	Temperature		Hovering Ceiling Hp - Ft.	
			Dry Air	80% R. H.
1600	-20°F	-28.9°C	6700	6500
	20°F	-6.7°C	5500	5200
	60°F	15.6°C	4300	3900
	100°F	37.8°C	3000	1300
1500	-20°F	-28.9°C	8100	7900
	20°F	-6.7°C	7100	6800
	60°F	15.6°C	5900	5600
	100°F	37.8°C	4800	2900
1400	-20°F	-28.9°C	9900	9700
	20°F	-6.7°C	8700	8400
	60°F	15.6°C	7400	7100
	100°F	37.8°C	6300	4400
1300	-20°F	-28.9°C	11700	11400
	20°F	-6.7°C	10400	10100
	60°F	15.6°C	9400	9000
	100°F	37.8°C	8200	6100

Courtesy Brantly Helicopter Corporation—

FIGURE 56.—Hovering ceiling (in ground effect) chart.

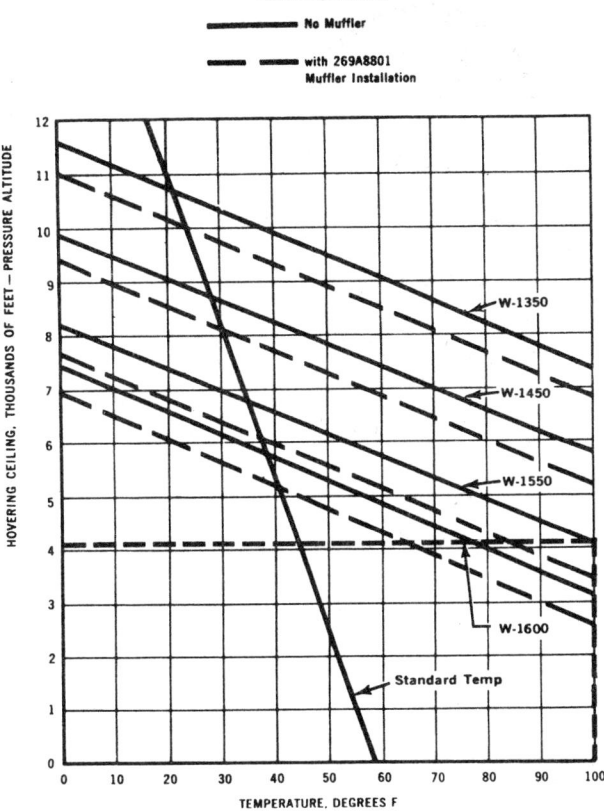

FIGURE 57.—Hovering ceiling (in ground effect) chart.

Rate of climb

For any given gross weight, the higher the density altitude, the less the rate of climb for any helicopter. Although a helicopter may be able to take off and clear obstacles close by, higher obstacles farther away may not be cleared because of this reduced rate of climb.

Figure 59 illustrates the type of chart that is used to compute the rate of climb for one model of helicopter. The steps for using this chart are exactly the same as those listed previously for the takeoff distance chart.

Figure 60 illustrates another type of chart used to compute rate of climb for another model helicopter. To use this chart, you must follow these steps:

1. Compute the density altitude at the departure point using the density altitude chart (figure 51).

2. Locate this density altitude along the left side of the chart (figure 60).

3. Follow this altitude line horizontally until it intersects the diagonal line.

4. From this point of intersection, move vertically downward to the bottom of the chart where you read the rate of climb under the existing conditions.

From the chart to the right in figure 60, you can determine the best rate of climb airspeed under the various density altitude conditions. From the density altitude at the left side of the chart, move horizontally across until the line intersects the diagonal line on the right-hand chart. From this point of intersection, move vertically downward to the bottom of the chart where you will read the best rate of climb speed in miles per hour.

Landing

Because a pilot can hover at his takeoff point with a certain gross weight, it does not mean that he can hover at his destination. If his destination is at a higher altitude and/or a higher temperature and moisture content prevail, he may not have sufficient power available to hover at his destination gross weight. The pilot will have to make a running landing under these conditions.

The pilot should be able to predict whether he will have sufficient power to hover at his destination through (1) a knowledge of temperature, relative humidity, and wind conditions at his destination; (2) the use of charts in the helicopter flight manual such as figures 56 and 57; and (3) by making certain power checks in flight prior to attempting to land. These power checks will be discussed later.

Figure 61 illustrates the type of chart that is used to compute the total landing distance over a 50-foot obstacle for one model of helicopter. The steps for using this chart are exactly the same as those listed previously for the takeoff distance chart. Notice that the total landing distance does not vary as much as the total takeoff distance (fig. 58) with equal variations in gross weight, altitude, or temperature. These charts apply to the same helicopter.

EFFECT OF GROSS WEIGHT ON HELICOPTER PERFORMANCE

We learned earlier that the total weight of a helicopter is the first force that must be overcome before flight is possible. Lift is the force that is needed to overcome or balance this total weight. It is easily seen that the greater the gross weight of the helicopter, the more lift that is required to hover. The amount of lift available is dependent upon the angle of attack at which the rotor blades can operate and still maintain required rotor RPM. The angle of attack at which the blades can operate at required rotor RPM is dependent upon the amount of power available. Therefore, the heavier the gross weight, the greater the power required to hover and for flight in general, and the poorer the performance of the helicopter since less reserve power is available (fig. 62); or, to state it another way, the heavier the gross weight, the lower the hovering ceiling.

		TAKE-OFF DISTANCE-FEET TO CLEAR 50 FOOT OBSTACLE			
		AT 50 MPH	3200 RPM		
Gross Weight Pounds	Pressure Altitude Feet	At -13°F -25°C	At 23°F -5°C	At 59°F 15°C	At 95°F 35°C
2150	SL	373	401	430	458
	2000	400	434	461	491
	4000	428	462	494	527
	6000	461	510	585	677
	8000	567	674	779	896
2500	SL	531	569	613	652
	2000	568	614	660	701
	4000	611	660	709	759
	6000	654	727	848	986
	8000	811	975	1144	1355
2850	SL	743	806	864	929
	2000	770	876	929	1011
	4000	861	940	1017	1102
	6000	939	1064	1255	1538
	8000	1201	1527	-	-

———— Courtesy Bell Helicopter Corporation ——

FIGURE 58.—Takeoff distance chart.

A study of the hovering ceiling chart (fig. 56) reveals the following interesting information for one helicopter, and this is fairly typical for all helicopters (with unsupercharged engines). At 60° F. in dry air, the hovering ceilings for gross weights of 1,300, 1,400, 1,500, and 1,600 pounds are 9,400 feet, 7,400 feet, 5,900 feet, and 4,300 feet, respectively. An increase of 300 pounds in gross weight decreases the hovering ceiling by more than half. At gross weights of 1,300, 1,400, 1,500, and 1,600 pounds at a temperature of 100° F. in air with a relative humidity of 80 percent, the hovering ceilings are 6,100 feet, 4,400 feet, 2,900 feet, and 1,300 feet, respectively. In the latter case, an increase of 300 pounds in gross weight reduces the hovering ceiling by almost 80 percent. *A comparison of the two examples illustrates vividly the reduction in performance brought about by a combination of heavy gross weight and high density altitudes.*

Of the three major factors affecting the performance of a helicopter at high elevations (density altitude, wind, and gross weight), the pilot can control only the gross weight. It should be obvious that the gross weight carried on any flight must be considered—not only for takeoff under the existing density altitude, wind conditions, and power available at point of departure, but also under the expected density altitude, wind conditions, and power available at the landing destination. Smaller amounts of fuel may be carried to improve performance or to increase useful load. It must be remembered, however, that this necessitates a sacrifice in range.

The importance of loading a helicopter within the approved center-of-gravity limits and the ill effects on performance if this is not properly accomplished, have been discussed in the preceding chapter.

EFFECT OF WIND ON HELICOPTER PERFORMANCE

We have seen earlier that when the horizontal airspeed of the helicopter reaches approximately 15 miles per hour, an abrupt increase in lift is experienced. This we called effective translational lift. Actually, from the moment the helicopter begins forward flight, translational lift is present, but is not very apparent or effective under about 15 miles per hour.

R/C MAX.   MAXIMUM RATE OF CLIMB FEET PER MINUTE   AT 50 MPH-3200 RPM

Gross Weight Pounds	Pressure Altitude Feet	At -25°C -13°F		At -5°C 23°F		At 15°C 59°F		At 35°C 95°F	
		(1)	(2)	(1)	(2)	(1)	(2)	(1)	(2)
2150	SL	1095	1235	1065	1210	1040	1190	1015	1145
	2,000	1065	1220	1040	1155	1015	1065	985	-
	4,000	1040	1085	995	-	910	-	830	-
	6,000	925	-	835	-	750	-	675	-
	8,000	770	-	685	-	600	-	530	-
	10,000	615	-	535	-	455	-	380	-
2500	SL	805	945	780	915	750	885	730	850
	2,000	780	915	750	855	725	775	705	-
	4,000	755	800	710	-	630	-	560	-
	6,000	650	-	565	-	485	-	410	-
	8,000	505	-	420	-	340	-	265	-
	10,000	360	-	275	-	195	-	120	-
2850	SL	560	685	535	660	500	625	470	580
	2,000	535	660	505	600	470	515	435	-
	4,000	505	545	455	-	375	-	295	-
	6,000	400	-	315	-	235	-	155	-
	8,000	260	-	175	-	95	-	15	-
	10,000	115	-	35	-	-	-	-	-

NOTE:

(1)   Continuous Power
(2)   Two Minute Power Rating

——Courtesy Bell Helicopter Corporation——

FIGURE 59.—Maximum rate of climb chart.

Translational lift is created by airspeed, not groundspeed. Therefore, translational lift is also present when the helicopter is hovering in a wind. If the wind velocity is 15 miles per hour or more, the helicopter will be experiencing effective translational lift in a hover. Due to this increased lift, less power will be required to hover than would be required for hovering in a no-wind condition (fig. 63); or, a greater gross weight could be carried when takeoff is to be made in a wind exceeding 15 miles per hour than could be carried if takeoff is to be made in a no-wind condition.

No-wind conditions increase the amount of power necessary to hover or require that a lighter load be carried. *Thus, no-wind conditions reduce helicopter performance.* Since wind decreases the power required for hovering or permits taking off or landing with greater loads, helicopter performance is improved. If the wind exceeds 15 miles per hour, performance is improved considerably; however, wind gusts over 30 to 35 miles per hour may tend to destroy the additional lift obtained between 15 and 30 miles per hour.

PRACTICAL METHODS FOR PREDICTING HELICOPTER PERFORMANCE

Certain practical methods for predicting helicopter performance were developed through engineering and flight tests for a particular model helicopter used by the Army. These practical methods for this particular helicopter are given in this handbook *to give the reader a clearer understanding of factors influencing helicopter performance, and sound principles on which to base flight decisions.* We wish to emphasize the fact that these rules are for a particular helicopter used by the Army and the

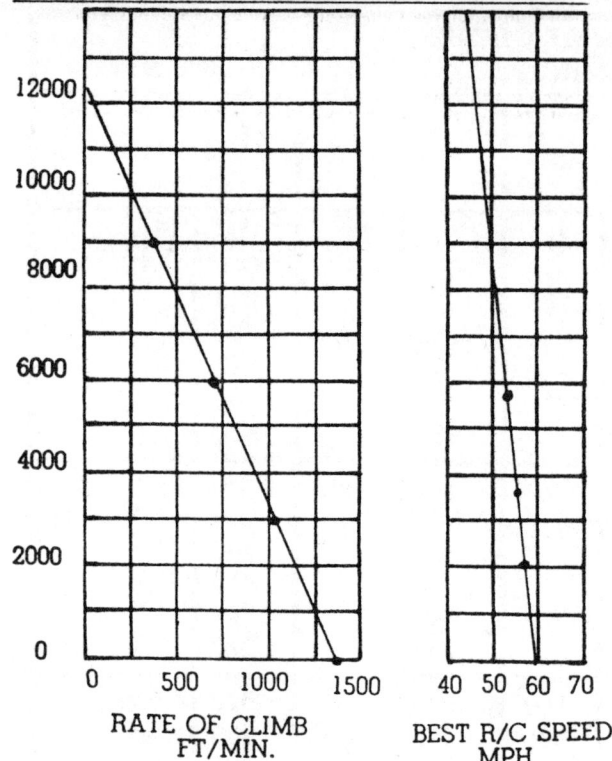

RATE OF CLIMB
FT/MIN.

BEST R/C SPEED
MPH

—————————————————Courtesy Brantly Helicopter Corporation—

FIGURE 60.—Rate of climb and best rate of climb speed chart.

actual figures will apply only to this particular helicopter. Even though such practical aids are developed for a helicopter, they should not be used as substitutes for experience and good judgment.

Manifold pressure and payload

Tests on this particular helicopter showed that 1 inch of manifold pressure was equivalent to 6 horsepower (hp), and that 1 hp would lift 13.5 pounds of weight while hovering. When combined, these two facts give rise to this practical rule:

Rule No. 1—One inch of manifold pressure will lift 80 pounds of payload.

With this knowledge, the pilot can obtain an estimate of the additional weight he can safely carry to be able to hover and then to enter flight. This rule may be applied before landing at destination in this manner:

1. Momentarily apply full throttle at 100 feet, or less, above the ground and determine the maximum manifold pressure that can be obtained. This will be approximately equal to the maximum manifold pressure available for takeoff.

2. While hovering, check manifold pressure required for the hover.

3. Find the difference between maximum available manifold pressure and manifold pressure required to hover.

4. The difference in manifold pressure changed into its equivalent in weight (1 inch of manifold pressure is equivalent to 80 pounds) gives the approximate additional payload which can be carried to lift to a hover for safe takeoff.

Temperature, winds, altitude, and gross weight are included in the above practical method for this particular helicopter and need not be considered separately.

Manifold pressure and hovering ceiling

By using available manifold pressure to determine hovering ceiling, a pilot can predict whether or not he can hover at a destination.

Rule No. 2—If wind velocity at point of intended landing is approximately the same as at point of takeoff, and the flight is made within the same air mass (no radical temperature change), for each inch of manifold pressure in excess of that required to hover, add 1,000 feet to the point-of-takeoff altitude. This computed altitude will represent the approximate hovering ceiling.

This practical rule may be applied as follows:

1. Check manifold pressure at a normal hover prior to departure.

2. While hovering, momentarily apply full throttle and note the maximum manifold pressure available.

3. The difference in these two manifold pressure readings is equivalent to 1,000 feet altitude per inch of excess manifold pressure. This additional altitude added to the point-of-takeoff altitude will give the maximum altitude (above sea level) at which the helicopter may be hovered (in ground effect).

Payload and wind

In winds from 0 to about 15 miles per hour, the hovering ceiling of the helicopter will increase about 100 feet for each mile per hour of wind. In winds from about 15 MPH to 26 MPH, the hovering ceiling will increase about 350 feet for each mile per hour of wind.

Rule No. 3—The payload may be increased 8 pounds for each mile per hour of wind from 0 to 15 miles per hour, or may be increased 28 pounds for each mile per hour of wind from 15 mph to 26 mph.

Hovering and skid height

The hovering altitude over level terrain for this particular helicopter is ideal with a skid clearance of approximately 4 feet (height of skid above the ground). Variable hovering altitudes, due to obstacles or rough terrain, have a decided effect on helicopter performance in deter-

Gross Weight Pounds	Pressure Altitude Feet	At -25°C -13°F	At -5°C 23°F	At 15°C 59°F	At 35°C 95°F
		TOTAL LANDING DISTANCE IN FEET OVER 50 FOOT OBSTACLE POWER-OFF AT 50 MPH			
2150	SL	243	253	265	277
	2000	253	267	278	293
	4000	264	278	294	319
	6000	278	293	310	327
	8000	293	310	330	350
2500	SL	248	258	270	282
	2000	258	272	283	298
	4000	269	283	299	314
	6000	283	298	315	332
	8000	298	316	335	355
2850	SL	282	294	307	320
	2000	293	309	322	338
	4000	306	322	340	357
	6000	322	340	358	378
	8000	340	359	380	403

— Courtesy Bell Helicopter Corporation

FIGURE 61.—Landing distance chart.

mining hovering ceiling and payload. These effects are best estimated as follows:

Rule No. 4—(1) To hover under 4 feet, add 300 feet to the hovering ceiling or 24 pounds to the payload for each 6 inches of decrease in skid height from the 4-foot hover.

(2) To hover between 4 feet and 10 feet, subtract 300 feet from the hovering ceiling or 24 pounds from the payload for each foot of increase in skid height.

Hovering ceiling and gross weight

The hovering ceiling will vary in proportion to the gross weight of the helicopter. To determine hovering ceiling for a known gross weight, apply the following rule:

Rule No. 5—(1) A 100-pound REDUCTION in gross weight increases hovering ceiling in or out of ground effect about 1,300 feet.

(2) A 100-pound INCREASE in gross weight decreases hovering ceiling about 1,300 feet.

Service ceiling and gross weight

The service ceiling of the helicopter varies with gross weight. (For all practical purposes, service ceiling is the maximum obtainable altitude.) To determine the effects of gross weight on service ceiling, apply the following rule:

Rule No. 6—A 100-pound DECREASE in gross weight adds 800 feet to the service ceiling, and, conversely, a 100-pound INCREASE in gross weight reduces the service ceiling 800 feet.

We wish to re-emphasize that these rules are for one particular helicopter used by the Army and the actual figures will apply only to this helicopter.

BRIEF SUMMARY

A thumbnail summary of this chapter might be as follows:

1. The most favorable conditions for helicopter performance are the combination of a low-density altitude, light gross weight, and moderate to strong wind (fig. 64).

2. The most adverse conditions for helicopter performance are the combination of a high-density altitude, heavy gross weight, and calm or no wind (fig. 64).

3. Any other combination of density altitude, gross weight, and wind conditions falls somewhere between the most adverse conditions and the most favorable conditions.

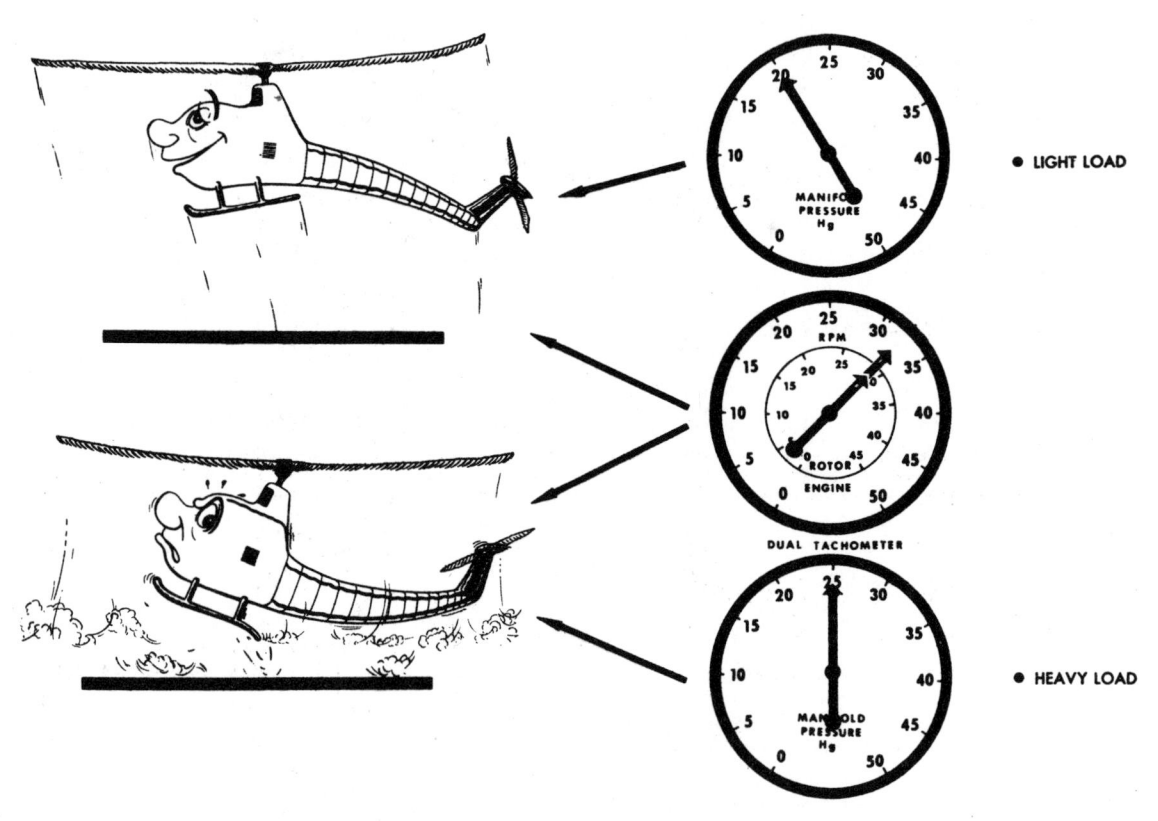

LIGHT LOAD

HEAVY LOAD

FIGURE 62.—Heavy loads (high gross weights) decrease helicopter performance.

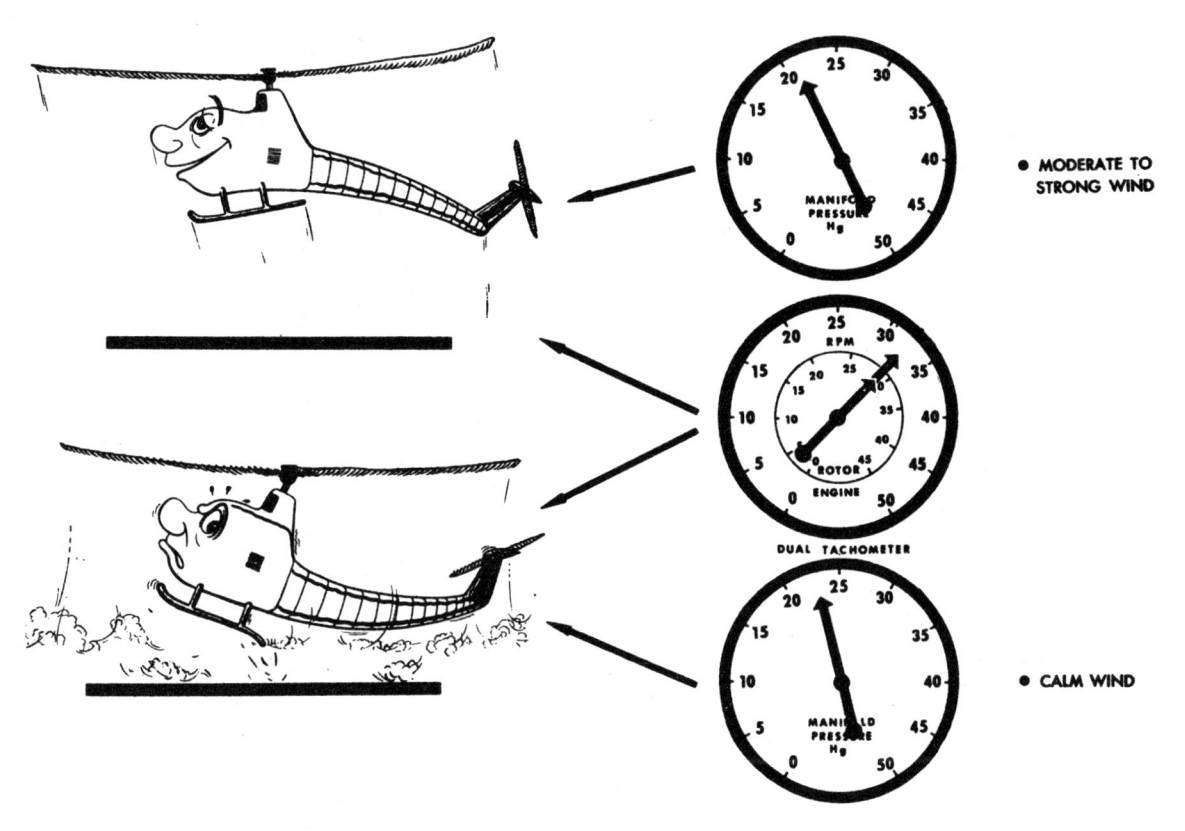

FIGURE 63.—Calm wind reduces helicopter hovering performance.

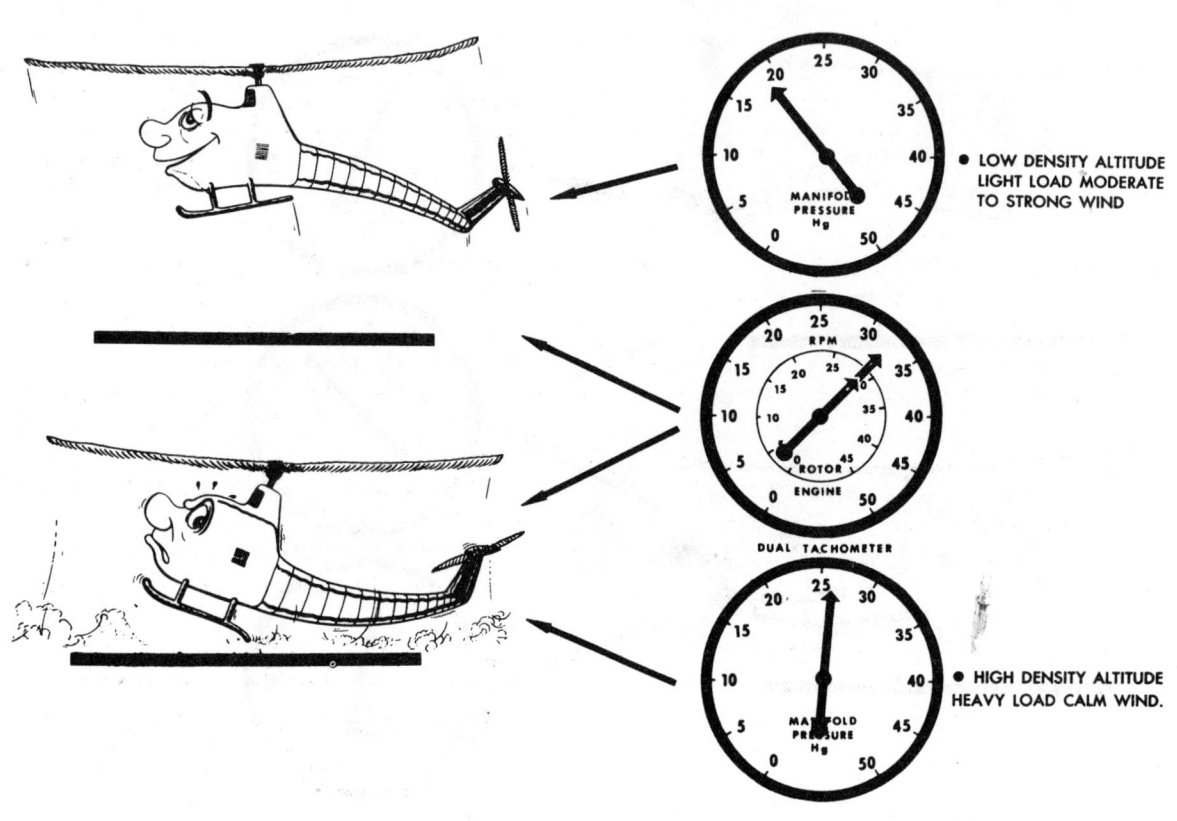

FIGURE 64.—The most adverse conditions for helicopter performance include the combination of high density altitude, heavy load (high gross weight) and calm wind.

Chapter 9. SOME HAZARDS OF HELICOPTER FLIGHT

Retreating blade stall

A tendency for the retreating blade to stall in forward flight is inherent in all present-day helicopters, and is a major factor in limiting their forward airspeed. Basically, the stall of the wing limits the low airspeed capabilities of the airplane. *The stall of a rotor blade limits the high airspeed potential of a helicopter.* The airflow over the retreating blade of the helicopter slows down as forward airspeed of the helicopter increases; the airflow over the advancing blade speeds up as forward airspeed increases. The retreating blade must, however, produce the same amount of lift as the advancing blade. Therefore, as the airflow over the retreating blade decreases with forward airspeed, the blade angle of attack must be increased to help equalize lift throughout the rotor disc area. As this increase in angle of attack is continued, the retreating blade will stall at some high forward airspeed. The advancing blade has relatively low angles of attack and is not subject to blade stall. Blade stall occurs during powered flight at the tip of the retreating blade, spreading inboard as forward airspeed increases. Retreating blade stall does not occur in normal autorotations.

When operating at *high forward airspeeds,* the following conditions are most likely to produce blade stall:

1—High gross weight.
2—Low RPM.
3—High density altitude.
4—Steep or abrupt turns.
5—Turbulent air.

The major warnings of approaching retreating blade stall conditions in the order in which they will generally be experienced are:

1—Abnormal 2 per revolution vibration in two-bladed rotors or 3 per revolution vibration in three-bladed rotors.
2—Pitchup of the nose.
3—Tendency for the helicopter to roll in the direction of the stalled (left) side.

At the onset of blade stall vibration, the pilot should take the following corrective measures:

1—Reduce collective pitch.
2—Increase rotor RPM.

3—Reduce forward airspeed.
4—Minimize maneuvering.

When operating under flight conditions likely to produce blade stall, a helicopter may quickly advance into severe blade stall by a steep turn, pullup, or other abrupt maneuver. The stall reaction will be rapid and violent. The vibrations, pitchup, and roll tendencies of the helicopter will present a serious threat to helicopter control and structural limitations. When flight conditions are such that blade stall is likely, extreme caution should be exercised when maneuvering.

As the altitude increases, the never-exceed airspeed (red line) for most helicopters decreases. Figure 65 shows a chart from the helicopter flight manual for one model from which the V_{ne} (never exceed) speed can be determined for the various altitudes. At sea level, V_{ne} is 86 miles per hour; at 6,000 feet and 2500 RPM, it is 65 MPH; and at 6,000 feet and 2700–2900 RPM, it is 78 MPH. This chart immediately points up the effect that rotor RPM has on the airspeed at which retreating blade stall is experienced.

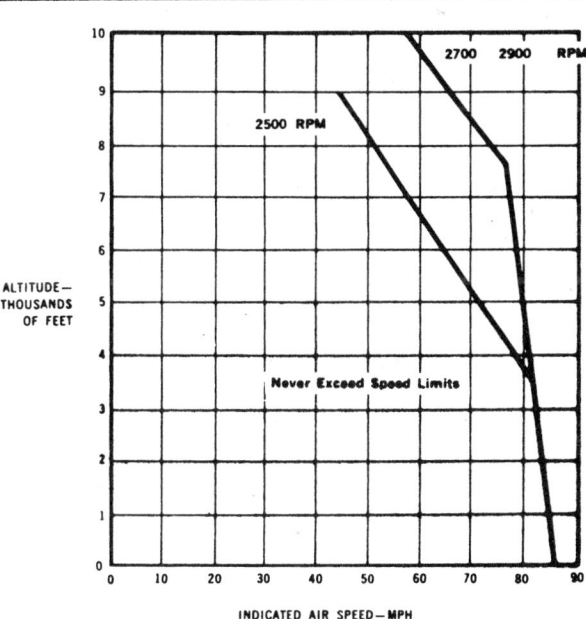

——————— Courtesy Hughes Tool Company, Aircraft Division ———

FIGURE 65.—Chart showing never-exceed (V_{ne}) speed limits.

Settling with power

This condition of flight is sometimes described as settling in your own downwash. It involves high vertical rates of descent and the addition of more power produces an even greater rate of descent. The helicopter is descending in turbulent air that has just been accelerated downward by the rotor. Reaction of this air on rotor blades at high angles of attack stalls the blades at the hub (center of the rotor) and the stall progresses outward along the blade as the rate of descent increases. The following combination of conditions are likely to cause settling with power:

1—A vertical or nearly vertical descent of at least 300 feet per minute. Actual critical rate depends on the gross weight, RPM, density altitude, and other pertinent factors.

2—The rotor system must be using some of the available engine power (from 20 to 100 percent).

3—The horizontal velocity must be no greater than approximately 10 miles per hour.

A pilot may experience settling with power accidentally. Situations that are conducive to a settling-with-power condition are:

1—Attempting to hover out of ground effect at altitudes above the hovering ceiling of the helicopter;

2—Attempting to hover out of ground effect without maintaining precise altitude control; or

3—A steep power approach in which airspeed is permitted to drop nearly to zero.

In recovering from a settling-with-power condition, the tendency on the part of the pilot to first try to stop the descent by increasing collective pitch will result in increasing the stalled area of the rotor and increasing the rate of descent. Since inboard portions of the blades are stalled, cyclic control will be reduced. Recovery can be accomplished by increasing forward speed, and/or partially lowering collective pitch.

Ground resonance

Ground resonance may develop when a series of shocks cause the rotor head to become unbalanced. This condition, if allowed to progress, can be extremely dangerous and usually results in structural failure. In general, if ground resonance occurs, it will occur only in helicopters possessing three-bladed, fully articulated rotor systems and using landing wheels. The rotor blades in a three-bladed helicopter are equally spaced around the rotor hub (120° apart), but are constructed to allow some horizontal movement. This horizontal movement is called drag and the vertical hinge that makes this possible is the drag hinge.

As the name implies, ground resonance occurs when the helicopter makes contact with the ground during landing or while in contact with the ground during an attempted takeoff. When one wheel of the helicopter strikes the ground first, a shock is transmitted through the fuselage to the rotor. This shock may cause the blades straddling the contact point to be forced closer together. The spacing might then be 122°, 122°, and 116°. When one of the other wheels strikes, the unbalance could be aggravated and become even greater. This establishes a resonance which sets up a pendulumlike oscillation of the the fuselage—a severe wobbling or shaking similar to the oscillations of a silver dollar, or similar object, when dropped striking the floor at an angle. Unless immediate corrective action is taken, the oscillations will increase rapidly and destruction of the helicopter will result. Corrective action could be an immediate takeoff if RPM is in proper range, or an immediate closing of the throttle and placing the blades in low pitch if RPM is low.

ABNORMAL VIBRATIONS

Abnormal vibrations in the helicopter will generally fall into three ranges:

1—Low frequency—100 to 400 cycles per minute (cpm).

2—Medium frequency—1,000 to 2,000 cpm.

3—High frequency—2,000 cpm or higher.

Low-frequency vibrations

Abnormal vibrations in this category are always associated with the main rotor. The vibration will be some frequency related to the rotor RPM and the number of blades of the rotor such as one vibration per revolution (1 per rev.), 2 per rev., or 3 per rev. Low-frequency vibrations are slow enough that they can be counted.

The frequency and the strength of the vibration will cause the pilot or passengers to be bounced or shaken noticeably. If the vibration is felt through the cyclic stick, it will have the same definite kick at the same point in the cycle. These low-frequency vibrations may be felt only in the fuselage or only in the stick or they may be evident in both at the same time. Whether the tremor is in the fuselage or the stick will, to some extent, determine the cause.

Those vibrations felt through the fuselage may be classified in three ways—lateral, longitudinal, or vertical—or they may be some combination of the three. A lateral vibration is one which throws the pilot from side to side. A longitudinal vibration is one which throws the pilot forward and backward, or in which the pilot receives a periodic kick in the back. A vertical vibration is one in which the pilot is bounced up and down, or it may be thought of as one in which the pilot receives a periodic kick in the seat of the pants. Describing the vibrations to the mechanic in this way will also help him determine the cause.

If the vibration is felt definitely in both the stick and fuselage, the cause is generally in the rotor or the rotor support. A failure of the pylon support at the fuselage is also a possible cause.

If the low-frequency vibration in the fuselage occurs only during translational flight or during a climb at certain airspeeds, the vibration may be a result of the blades striking the blade rest stops. This can be eliminated by avoiding the flight condition that causes it.

For low-frequency vibrations felt predominantly through the stick, the most likely place to look for trouble is in the control system linkage from the stick to the rotor head.

Medium-frequency vibrations

Medium-frequency vibrations are a result of trouble with the tail rotor in most helicopters. Improper rigging, unbalance, defective blades, or bad bearings in the tail rotor are all sources of these vibrations. If the vibration occurs only during turns, the trouble may be caused by insufficient tail rotor flapping action. Medium-frequency vibrations will be very difficult if not impossible to count due to the fast rate.

High-frequency vibrations

High-frequency vibrations are associated with the engine in most helicopters, and will be impossible to count due to the high rate. However, they could be associated with the tail rotor for helicopters in which the tail rotor RPM is approximately equal to or greater than the engine RPM. A defective clutch or missing or bent fan blades will cause vibrations which should be corrected. Any bearings in the engine or in the transmission or the tail rotor drive shaft that go bad will result in vibrations with frequencies directly related to the speed of the engine.

Experience in detecting and isolating the three different classes of vibrations when they first develop makes it possible to correct the vibrations long before they become serious.

TRANSITION FROM POWERED FLIGHT TO AUTOROTATION

It is obvious that there are some transitions that are necessary in establishing and stabilizing a helicopter in autorotative flight following a power failure. A power failure immediately results in a dissipation of rotor RPM while the airflow through the rotor system is changing from the downward flow present in powered flight to the upward flow occurring during autorotative flight. Rotor RPM decreases at a rapid rate if immediate action is not

taken to decrease the pitch angle of the rotor blades. This decrease continues (even after the upward flow of air has stabilized) to the point that controlled flight may not even be possible depending upon the pitch angle at the time of power failure. After the pitch angle is lessened and the upward flow of air stabilizes, the rate of descent will not stabilize at its minimum until the rotor RPM builds back up to its maximum for that particular pitch angle setting and helicopter gross weight.

The successful entry from powered flight to autorotation consists of the following transitions:

1—Changing of airflow from a downward flow to an upward flow.

2—Lowering collective pitch to maintain a tolerable angle of attack which would otherwise increase because of the descent.

3—Regaining rotor RPM and stabilizing rate of descent.

It is well to note that the magnitude of these transitions depends on the mode of flight. If power failure occurs in a descent, there is very little transition. A vertical climb requires large transitions because of the helicopter's upward inertia and the high rotor blade pitch angle required for the vertical climb.

Altitude versus airspeed charts

Figure 66 is an altitude versus airspeed chart excerpted from the helicopter flight manual for one helicopter. This type of chart is often referred to as the "height-velocity curve" or diagram and "dead man's curve." Such charts are prepared by the manufacturer and, as required by regulations, are published in the helicopter flight manual generally under the performance section.

A helicopter pilot must become familiar with this chart for the particular helicopter that he flies. From it, he is able to determine at what altitudes and airspeeds he can safely make an autorotative landing in case of an engine failure; or, to restate it in another way, he is able to determine those altitude-airspeed combinations from which it would be nearly impossible to successfully complete an autorotation landing. The altitude–airspeed combinations that should be avoided are represented by the shaded areas of the chart.

Imagine, if you will, the difficulty of entering and establishing a stabilized autorotation from the shaded areas of the "dead man's curve" (fig. 66). This chart is assuming the lack of vertical velocity and acceleration (that is, flight at a constant altitude and airspeed) and that an engine failure in these areas is complete and instantaneous. In either of these shaded areas, it would be nearly impossible to complete all the transitions to autorotation before the helicopter would be at ground level. When the heli-

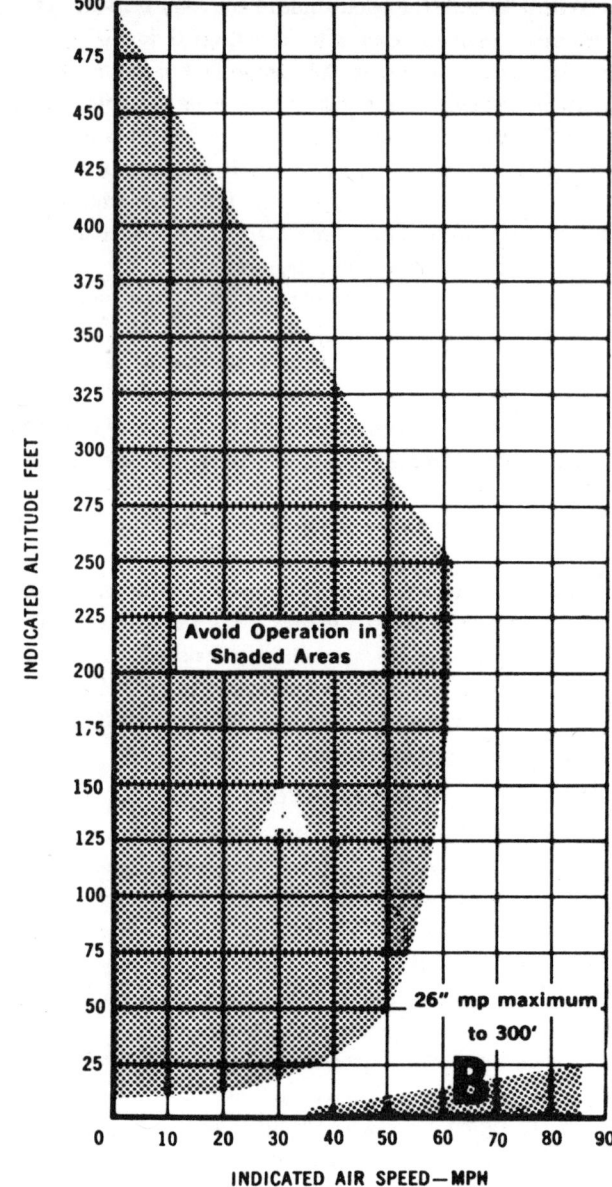

FIGURE 66.—Airspeed vs. altitude limitations chart.

copter is "pushed over" (that is, when the nose is lowered) into a glide, the rotor system angle of attack is diminished. This causes the helicopter to accelerate downward and lengthens the time required to complete the autorotational transitions. That is, the time required for the helicopter airspeed, rotor RPM, and rate of descent to stabilize would be greater than if the helicopter were already in the autorotative descending attitude with best autorotational airspeed at the time of engine failure.

In case of engine failure in the low altitude, low airspeed area of figure 66 (area A), a choice must be made by the pilot either to "freeze" the collective pitch to save the remaining rotor RPM and upward collective travel for cushioning the ground contact, or to lower the collective pitch all the way, or to partially lower the collective pitch so that more upward collective travel will be available to cushion the ground contact. It is strictly a judgment decision on the part of the pilot based on his skill, experience, and knowledge of his helicopter and the altitude-airspeed combination at the time of engine failure. In any case, the rate of descent at ground contact will be much greater than it would be from a normal autorotation in which all transitions could be made in time for the rate of descent to stabilize.

In the low altitude, high airspeed area of figure 66 (area B), the pilot's immediate choice is to flare or lower collective pitch or to decide how much of each may be done. The type of landing area would be a factor in this instance; however, even though a smooth landing area of sufficient length is available, the pilot still risks a high-speed landing in an aircraft which is not stressed for it.

Ascent of the helicopter in area A and forward acceleration in area B make the occurrence of a power failure doubly hazardous because of the magnitude of the induced airflow and collective pitch transitions required to enter a stabilized autorotation. A glide through either area is much less hazardous in case of power failure because most of the transitions have already been accomplished.

ANTITORQUE SYSTEM FAILURE

Antitorque system failure could be the result of a failure in the tail rotor blades, a failure in the mechanical linkage between the pedals and the pitch-change mechanism of the tail rotor, or a failure in the tail rotor drive shaft between the transmission and the tail rotor.

Antitorque system failure in forward cruising flight

If the antitorque system fails in forward cruising flight, the nose of the helicopter will usually pitch slightly and yaw to the right. The direction in which the nose will pitch depends on the particular helicopter and how it is loaded. Violence of the pitching and yawing is generally greater when the failure occurs in the tail rotor blades and is usually accompanied by severe vibration.

The pitching and yawing can be overcome by holding the cyclic stick near neutral and entering autorotation. Cyclic control movements should be kept to a minimum until all pitching subsides. Avoid abrupt rearward movements of the cyclic stick. If it is moved rearward abruptly, the main rotor blades could flex downward with sufficient force to strike the tail boom. If unsuitable terrain exists, cautiously add power to determine if flight can be continued to an acceptable landing area. If dangerous attitudes are incurred due to this addition of power, re-enter autorotation.

If sufficient forward speed is maintained, the fuselage remains fairly well streamlined; however, if descent is attempted at slow speeds, a continuous turning movement to the left can be expected. (Know the manufacturer's recommendations in case of tail rotor failure for each particular helicopter you fly. This will generally be found under Emergency Procedures in the helicopter flight manual.) Directional control should be maintained primarily with cyclic control and, secondarily, by gently applying throttle momentarily, with needles joined, to swing the nose to the right.

A landing may be made with forward speed or by flaring. The best and safest landing technique, terrain permitting, is to land directly into the wind with approximately 20 miles per hour airspeed. The helicopter will turn to the left during the flare and during the subsequent vertical descent. An important factor to remember is that the helicopter should be level or approximately level at ground contact.

Antitorque system failure while hovering

If the antitorque system fails during hovering flight, quick action must be taken by the pilot. The turning motion to the right builds up rapidly because of the torque reaction produced by the relatively high-power setting. The throttle should be closed immediately (without varying collective pitch position) to eliminate this turning effect. Simultaneously, the cyclic stick should be adjusted to stop all sideward or rearward movements and to level the helicopter for touchdown. From this point, the procedure for a hovering autorotation (p. 88) should be followed.

Chapter 10. PRECAUTIONARY MEASURES AND CRITICAL CONDITIONS

Because of its unique flight characteristics, a helicopter is capable of many missions no other aircraft can perform. A helicopter pilot must, however, realize the hazards involved and know what precautions to take that may save his helicopter or even his life.

General precautionary rules

1. Do not perform acrobatic maneuvers.
2. Do not check magnetos in flight.
3. Use caution when adjusting mixture in flight.
4. Always taxi slowly.
5. Always check ballast prior to flying.
6. Use caution when hovering on the leeward side of buildings or obstructions.
7. Do not hover at an altitude that will place you in the shaded area of the height–velocity chart.
8. Always hover for a moment before beginning a new flight.
9. When flying in rough, gusty air, use special care to maintain proper RPM.
10. When practicing hovering turns, sideward flight, and similar low airspeed maneuvers, be especially careful to maintain proper RPM.
11. Always clear the area overhead, ahead, to each side, and below before entering practice autorotations.
12. Make sure any object placed in the cockpit of a helicopter is secured to prevent fouling of the controls.
13. Except in sideward or rearward flight, always fly the helicopter from references ahead.

Rotor RPM operating limits

Limits of rotor RPM vary with each type of helicopter. In general, the lower limit is determined primarily by the control characteristics of the helicopter during autorotation. Since the tail rotor is driven by the main rotor, a minimum main rotor RPM exists at which tail rotor thrust is sufficient for proper heading control. Below this minimum main rotor RPM, full pedal pressure will not maintain heading control under certain conditions of flight.

The upper limit for rotor RPM is based on both autorotative characteristics and structural strength of the rotor system. Structural tests plus an adequate margin for safety are required by FAA safety standards for the certification of aircraft.

Extreme attitudes and overcontrolling

Design characteristics of a helicopter preclude the possibility of safe inverted flight. Avoid all maneuvers which would place a helicopter in danger of such an extreme attitude.

Avoid loading a helicopter so as to cause an extreme tail-low attitude when taking off to a hover. Aft center of gravity is dangerous while hovering and even more dangerous while in flight because of limited forward cyclic stick travel.

Avoid heavy loading forward of the center of gravity. The result is limited aft cyclic stick travel endangering controllability.

Avoid an extreme nose-low attitude when executing a normal takeoff. Such an attitude may require more power than the engine can deliver and will allow the helicopter to settle to the ground in an unsafe landing attitude. In the event of a forced landing, only a comparatively level attitude can assure a safe touchdown.

Avoid abrupt application of rearward cyclic control. The violent backward-pitching action of the rotor disc may cause the main rotor blades to flex downward into the tail boom.

Avoid large or unnecessary movements of the cyclic control while at a hover. Such movements of the cyclic control can, under certain conditions, cause sufficient loss of lift to make the helicopter settle to the ground.

Flight technique in hot weather

1. Make full use of wind and translational lift.
2. Hover as low as possible and no longer than necessary.
3. Maintain maximum allowable engine RPM.
4. Accelerate very slowly into forward flight.
5. Employ running takeoffs and landings when necessary.
6. Use caution in maximum performance takeoffs and steep approaches.
7. Avoid high rates of descent in all approaches.

Effect of altitude on instrument readings

The thinner air of higher altitudes causes the airspeed indicator to read "too low." True airspeed may be roughly computed by adding to the indicated airspeed, 2 percent of the indicated airspeed for each 1,000 feet of altitude above sea level. For example, an indicated airspeed of 80 MPH at 5,000 feet will be a true airspeed of approximately 88 MPH. This computation may be made more accurately by using a computer.

Manifold pressure is reduced approximately 1 inch for each 1,000 feet of increase in altitude. If a maximum manifold pressure of 28 inches can be obtained at an elevation of 1,000 feet, only 22 inches of manifold pressure will be available at 7,000 feet. This loss of manifold pressure must be considered when planning flights from low altitudes to high altitudes.

High altitude pilot technique

Of the three major factors limiting helicopter performance at high altitudes (gross weight, density altitude, and wind), only gross weight may be controlled by the pilot of a helicopter with an unsupercharged engine. At the expense of range, smaller amounts of fuel may be carried to improve performance or to increase the number of passengers or the amount of baggage. Where practical, running landings and takeoffs should be used. Make maximum use of favorable wind, with landings and takeoffs directly into the wind if possible.

When the wind blows over large obstructions such as mountain ridges, a condition is set up similar to that depicted in the upper right of figure 81; however, it is a much more disturbed condition. The wind blowing up the slope on the windward side is usually relatively smooth. However, on the leeward side the wind spills rapidly down the slope, similar to the way water flows

down a rough streambed, setting up strong downdrafts and causing the air to be very turbulent (fig. 67). The downdrafts can be very violent and may cause aircraft to strike the sides of the mountains. Therefore, when approaching mountain ridges against the wind, it is sound practice to make an extra altitude allowance to assure safe terrain clearance. Where pronounced mountain ridges and strong winds are present, a clearance of 2,000 or 3,000 feet above the terrain is considered a desirable minimum. Also, it is advisable to climb to the crossing altitude well before reaching the mountains to avoid having to make the climb in a persistent downdraft.

When operating over mountainous terrain, fly on the upwind side of slopes to take advantage of updrafts. When landing on ridges, the safest approach is usually made lengthwise of the ridge. Fly near the upwind edge to avoid possible downdrafts and to be in position to autorotate down the upwind side of the slope in case of a forced landing. "Riding" the updraft in this manner results in a lower rate of descent, improved glide ratio, and greater choice of a landing area.

Tall grass and water operations

Tall grass will tend to disperse or absorb the ground cushion. More power will be required to hover, and takeoff may be very difficult. Before attempting to hover over tall grass, make sure that at least 2 or 3 inches more manifold pressure is available than is required to hover over normal terrain.

Operations over water with a smooth or glassy surface renders accurate altitude determination very difficult.

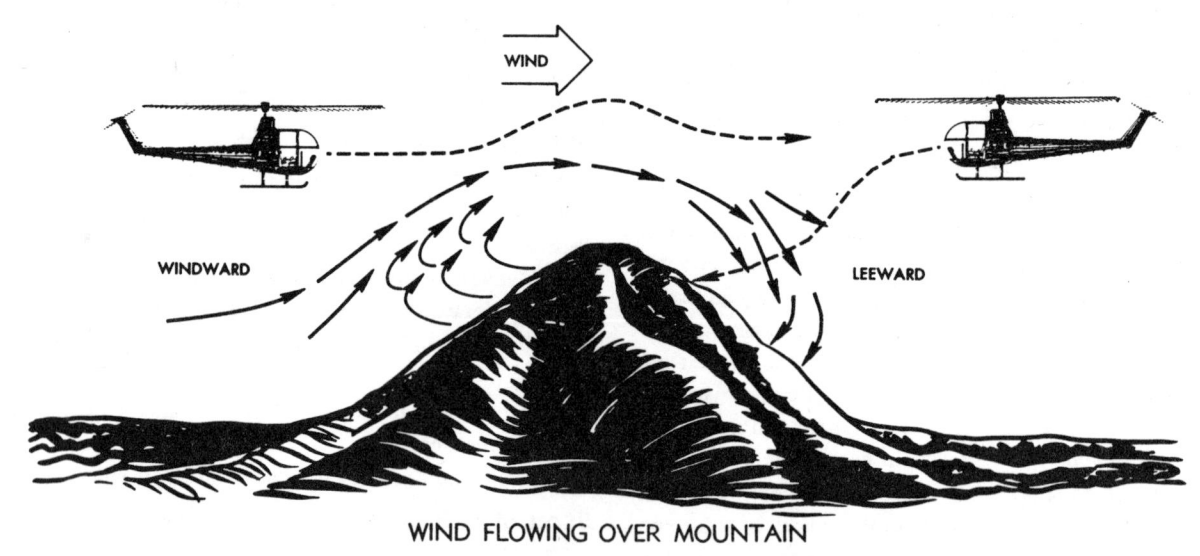

WIND FLOWING OVER MOUNTAIN

FIGURE 67.—Windflow over mountains and its effect on aircraft in flight.

Thus, caution must be exercised to prevent the helicopter from inadvertently striking the water. This problem does not exist over rough water but a very rough water surface may disperse the ground effect and thereby require more power to hover. Movements of the water surface, wind ripples, waves, currentflow, or even agitation by the helicopter's own rotor wash tend to give the pilot a false feeling of helicopter movement.

CARBURETOR ICING

Carburetor icing is a frequent cause of engine failure. The vaporization of fuel, combined with the expansion of air as it passes through the carburetor, causes a sudden cooling of the mixture. The temperature of the air passing through the carburetor may drop as much as 60° F. within a fraction of a second. Water vapor in the air is "squeezed out" by this cooling, and if the temperature in the carburetor reaches 32° F. or below, the moisture will be deposited as frost or ice inside the carburetor passages. Even a slight accumulation of this deposit will reduce power and may lead to complete engine failure, particularly when the throttle is partly or fully closed.

Conditions favorable for carburetor icing

On dry days, or when the temperature is well below freezing, the moisture in the air is not generally enough to cause trouble. But if the temperature is between 20° F. and 70° F., with visible moisture or high humidity, the pilot should be constantly on the alert for carburetor ice. During low or closed throttle settings, an engine is particularly susceptible to carburetor icing.

Indications of carburetor icing

Indications of carburetor ice include unexplained loss of RPM or manifold pressure; the carburetor air temperature indicating in the "danger" (red arc) or "caution" (yellow arc) range; and engine roughness. A loss of manifold pressure will generally give the first indication; however, due to the many small control (setting) changes made in the throttle and collective pitch, this may be less noticeable. Therefore, a close check of the carburetor air temperature gage is necessary so that carburetor heat may be adjusted to keep the carburetor air temperature gage out of the red and yellow arcs.

Carburetor air temperature gage

Carburetor air temperature gages are marked with a green arc representing the range of desired operating temperatures; a yellow arc representing the range of temperatures in which caution should be exercised since icing is possible; and a red line representing the maximum operating temperature limit. Sometimes a red arc is used to represent the most dangerous range in which carburetor ice can be anticipated. The carburetor heat control should be adjusted so that the carburetor air temperature remains in the green arc.

Use of carburetor heat

The carburetor heater is an anti-icing device that preheats the air before it reaches the carburetor. This preheating can be used to melt ice or snow entering the intake duct, to melt ice that forms in the carburetor passages (provided the accumulation is not too great); and to keep the fuel mixture above the freezing point to prevent formation of carburetor ice.

When conditions are favorable for carburetor icing, the pilot should make the proper check to see if any is present. He should check the manifold pressure gage reading, then apply full carburetor heat and leave it on until he is certain that if ice was present, it has been removed. (During this check a constant throttle and collective pitch setting should be maintained.) The carburetor heat should then be returned to the OFF (cold) position. If the manifold pressure gage indicates higher than when the check was initiated, carburetor ice was present.

When there are indications of carburetor icing, full carburetor heat should be applied until the manifold pressure returns to normal and the engine is running smoothly. The carburetor heat should then be adjusted so that the carburetor air temperature gage indicates a safe operating range.

Fuel injection

Fuel injection systems have replaced carburetors in some helicopters. In the fuel injection system, the fuel is normally injected into the system either directly into the cylinders or just ahead of the intake valve. In the carburetor system, the fuel enters the airstream at the throttle valve. The fuel injection system is generally considered to be less susceptible to icing than the carburetor system.

Chapter 11. HELICOPTER FLIGHT MANEUVERS

From the foregoing chapters in this handbook, it should be obvious that the variable factors of wind, temperature, humidity, gross weight, and structural differences of various helicopter models greatly affect the operation of the helicopter. Even when flying the same model helicopter, two flights are seldom exactly alike to the pilot because of variation in wind and density altitude. Therefore, it is practically impossible to prescribe helicopter attitudes for the performance of each flight maneuver since this handbook should generally apply to all helicopters having the characteristics set forth on page 1. Attitudes, airspeeds, altitudes, and power settings must of necessity vary slightly to suit the weather, the particular helicopter, and the loading. For example, on a day with a 20-MPH wind and a 1,000-foot density altitude, a certain nose-low attitude and power setting will be required to hover and to initiate a departure from a hover to commence a normal takeoff; the following day, with a no-wind condition and a 2,500-foot density altitude, both the nose attitude and the power setting for hovering and initiating a departure from the hover may differ considerably in degree from the previous day. Obviously, then, it would be impossible for the handbook to outline a specific nose attitude and power setting for departure from a hover. Therefore, this chapter does not detail each and every attitude of a helicopter in the various flight maneuvers nor each and every move a pilot must make in order to perform a given maneuver.

VERTICAL TAKEOFF TO A HOVER

A vertical takeoff (or takeoff to a hover) is a maneuver in which the helicopter is raised vertically from a spot on the ground to the normal hovering altitude with a minimum of lateral and/or fore and aft movement.

TECHNIQUE:

1. Head helicopter into the wind.
2. Place cyclic stick in neutral position.
3. Be sure that the collective pitch stick is in the full down position.
4. Increase the throttle smoothly to obtain and maintain proper hover RPM.
5. Raise the collective pitch. Use smooth, continuous movement, coordinating throttle to maintain proper RPM

setting. As collective pitch is increased and the helicopter becomes light on the skids, torque will tend to cause the nose to swing to the right unless the pilot adds a sufficient amount of left pedal pressure to maintain a constant heading.

6. As the helicopter becomes light on the skids, make necessary cyclic stick corrections to ensure a level attitude on becoming airborne, pedal corrections to maintain heading, and collective pitch corrections to ensure continuous vertical ascent to normal hovering altitude.

The higher the density altitude the lower the hovering altitude should be.

7. When hovering altitude is reached, adjust throttle and collective pitch as required to maintain proper RPM and altitude. Coordinate pedal changes with throttle and collective changes to maintain heading and use cyclic as necessary to maintain a constant position over the spot. Remember—collective pitch controls altitude, cyclic pitch controls attitude and position.

8. Check engine and control operation, manifold pressure required to hover, and note cyclic stick position. Cyclic stick position will vary with amount and distribution of load and wind velocity.

COMMON ERRORS:

1. Failing to maintain level attitude as the helicopter becomes airborne.
2. Pulling through on the collective after becoming airborne, causing the helicopter to gain too much altitude. This, in turn, necessitates comparatively large throttle and collective pitch changes.
3. Overcontrolling the pedals, which not only changes the heading of the helicopter but also changes RPM, thus necessitating constant throttle adjustment.
4. Reducing throttle rapidly in situations where proper RPM has been exceeded which usually results in violent changes of heading to the left and loss of lift resulting in loss of altitude.

HOVERING

Hovering is a maneuver in which the helicopter is maintained in nearly motionless flight over a reference point at a constant altitude and on a constant heading.

The maneuver requires a high degree of concentration and coordination on the part of the pilot. When hovering, a pilot holds the helicopter over a selected point by use of cyclic control; maintains altitude by use of collective pitch control; and maintains a constant heading with anti-pedals. Only by proper coordination of all controls can successful hovering flight be achieved.

Control corrections should be pressure rather than abrupt movements. A constant pressure on the desired pedal will result in a smooth rate of turn; pronounced movements tend to jerk the nose around. If the helicopter tends to move forward, a slight amount of back pressure on the cyclic control stick will stop the forward movement. Just before movement stops, back pressure must be released or the helicopter will come to a stop, and start into rearward flight. Avoid waiting out helicopter moves; make all corrections immediately. Stopping and stabilizing the helicopter at a hover requires a number of small pressure corrections to avoid over-controlling.

The attitude of the helicopter determines its movements over the ground. While the attitude required to hover varies with wind conditions and center-of-gravity location, there is a particular attitude which can be found by experimentation to keep the helicopter hovering over a selected point. After this attitude has been determined, deviations can be noted and necessary corrections made before the helicopter actually starts to move from the point.

Hovering altitude is maintained by use of collective pitch, coordinated with the throttle, to maintain a constant RPM. The amount of collective pitch needed to maintain hovering altitude varies with wind, air density (density altitude), and gross weight. When a steady wind is blowing, very little adjustment of the collective pitch stick should be required to hold a desired altitude. Only under variable and gusty wind conditions should any great collective pitch control changes be required.

Coordination of all controls cannot be overemphasized. Any change on one control will almost always require a coordinated correction on one or more of the other controls. Hovering can be accomplished in a precision manner only when corrections are small, smooth, and coordinated.

COMMON ERRORS:

1. Tenseness and late reactions to movements of the helicopter.
2. Failure to allow for lag in cyclic and collective pitch control which leads to overcontrolling.
3. Confusing altitude changes for attitude changes resulting in the use of improper cockpit control.
4. Hovering too high.
5. Hovering too low, resulting in occasional touchdown.

HOVERING TURN

A hovering turn is a maneuver performed at hovering altitude in which the nose of the helicopter is rotated left or right while maintaining position over a reference point on the ground. This maneuver requires the coordination of all flight controls and demands precision control near the ground. Constant altitude, rate of turn, and RPM should be maintained.

TECHNIQUE:

1. Initiate the maneuver from a normal hovering altitude headed into wind by starting a smooth application of pedal in the desired direction of turn.

2. As the nose begins to swing, and throughout the remainder of the turn, use cyclic control to maintain position over the ground reference point, use pedals to maintain a slow, constant rate of turn, and use collective pitch along with proper throttle coordination to maintain a constant altitude and proper operating RPM.

3. As the 180° position is approached, anticipate the use of a small amount of opposite pedal. As the tail of the helicopter swings from a position into the wind to a position downwind, the helicopter will have a tendency to whip or increase its rate of turn as a result of the weather-vaning tendency of the tail surface. The higher the winds the greater will be this whipping action.

4. As the desired heading on which the turn is to be completed is approached, apply opposite pedal pressure as necessary to stop the turn on this heading.

5. It should be noted that, during a hovering turn to the left, the RPM will decrease if throttle is not added; in a hovering turn to the right, RPM will increase if throttle is not reduced slightly. (This is due to the amount of engine power that is being absorbed by the tail rotor which is dependent upon the pitch angle at which the tail rotor blades are operating.) Avoid making large corrections in RPM while turning since the throttle adjustment will result in erratic nose movements due to torque changes.

6. To determine the amount of left pedal available, make the first hovering turn to the left. If a 90° turn to the left cannot be effected, or if an unusual amount of pedal is required to complete a 45° hovering turn to the left, do not attempt a turn to the right since sufficient left pedal may not be available to prevent an uncontrolled turn. Hovering power requires a large amount of left pedal to maintain heading. Sufficient left pedal *in excess of this amount* must be available to prevent an uncontrolled turn to the right once the turn has begun.

7. Hovering turns should be avoided in winds strong enough to preclude sufficient back cyclic control to maintain the helicopter on the selected ground reference point when headed downwind. Check the helicopter flight

manual for the manufacturer's recommendations for this limitation.

COMMON ERRORS:

1. Failing to maintain a slow, constant rate of turn.
2. Failing to maintain position over the reference point.
3. Failing to keep the RPM within normal operating range.
4. Failing to maintain constant altitude.
5. Failing to use pedals smoothly and cautiously.

HOVERING—FORWARD FLIGHT

Forward hovering flight can generally be used to move the helicopter to a specific area unless strong winds prohibit crosswind or downwind hovering. A hovering turn is utilized to head the helicopter in the direction of the desired area then forward flight at a slow speed is used to move to the area. During the maneuver, a constant groundspeed, altitude, and heading should be maintained.

TECHNIQUE:

1. Before starting forward flight, pick out two references directly in front of the helicopter and in line with it. These reference points should be kept in line throughout the maneuver (fig. 68).

2. Initiate the maneuver from a normal hovering altitude by applying forward pressure on the cyclic stick.

3. As movement begins, return the cyclic stick toward the neutral position to keep the groundspeed at a low rate—no faster than normal walking speed. At this speed, ground effect will be retained, thus reducing the need for power and pedal corrections.

4. Throughout the maneuver, maintain a constant groundspeed and ground track with cyclic stick, a con-

stant heading with pedals, a constant altitude with collective pitch control, and proper operating RPM with throttle.

5. To stop the forward movement, apply rearward cyclic pressure until the helicopter stops. As forward motion stops, the cyclic must be returned to the neutral position to prevent rearward movement. Forward movement can also be stopped by simply applying enough rearward cyclic pressure to level the helicopter and let it drift to a stop.

COMMON ERRORS:

1. Erratic movement of the cyclic stick, resulting in overcontrol and uneven movement over the ground.
2. Failure to use proper pedals properly, resulting in excessive nose movement.
3. Failure to maintain a hovering altitude.
4. Failure to maintain proper RPM.

HOVERING—SIDEWARD FLIGHT

Sideward hovering flight may be necessary to move the helicopter to a specific area when conditions make it impossible to use forward flight. During the maneuver, a constant groundspeed, altitude, and heading should be maintained.

TECHNIQUE:

1. Before starting sideward flight, pick out two reference points in a line in the direction sideward flight is to be made to help you maintain proper ground track (fig. 69). These reference points should be kept in line throughout the maneuver.

2. Initiate the maneuver from a normal hovering altitude by applying pressure on the cyclic stick toward the side in which movement is desired.

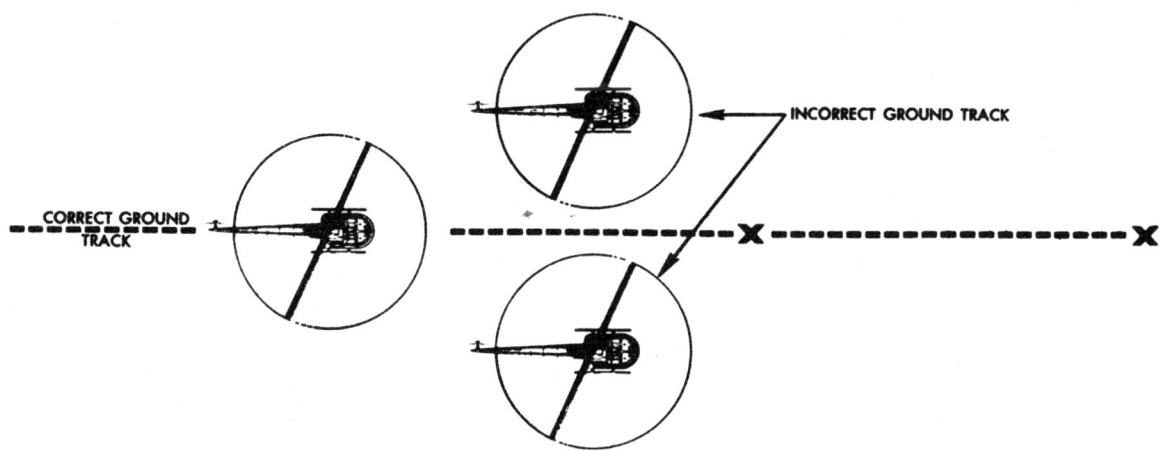

FIGURE 68.—Use of reference points in maintaining proper ground track in forward or rearward hovering flight.

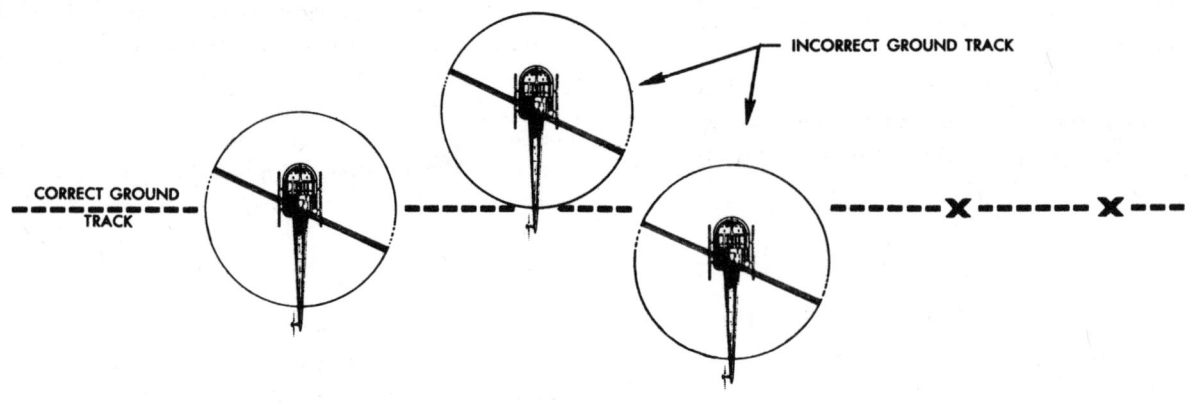

FIGURE 69.—Use of reference points in maintaining proper ground track in sideward hovering flight.

3. As movement begins, return the cyclic stick toward the neutral position to keep the groundspeed at a low rate—no faster than normal walking speed. At this speed, ground effect will be retained, reducing the need for power and pedal corrections.

4. Throughout the maneuver, maintain a constant groundspeed and ground track with cyclic stick, a constant heading perpendicular to the proposed ground track with pedals, a constant altitude with collective pitch control, and proper operating RPM with throttle.

5. To stop the sideward movement, apply cyclic pressure in the direction opposite to that of movement, and hold until the helicopter stops. As motion stops the cyclic stick must be returned to the neutral position to prevent movement in the opposite direction. Sideward movement also can be stopped by simply applying enough opposite cyclic pressure to level the helicopter. Then it will drift to a stop.

COMMON ERRORS:

1. Erratic movement of the cyclic stick, resulting in overcontrol and an uneven movement over the ground.

2. Failure to use proper antitorque control, resulting in excessive nose movement.

3. Failure to maintain a hovering altitude.

4. Failure to maintain proper RPM.

5. Failure to make clearing turns prior to starting the maneuver.

HOVERING—REARWARD FLIGHT

Rearward hovering flight may also be necessary at times to move the helicopter to a specific area when the situation is such that forward or sideward flight cannot be used. During the maneuver, a constant groundspeed, altitude, and heading should be maintained. The area behind the helicopter must be carefully cleared before the maneuver is begun.

TECHNIQUE:

1. Before starting rearward flight, pick out two reference points in front of, and in a line with the helicopter to help you maintain proper ground track (fig. 68). The movement of the helicopter should be such that these reference points remain in a line.

2. Initiate the maneuver from a normal hovering altitude by applying rearward pressure on the cyclic stick. After movement has begun, position the cyclic stick to maintain a slow groundspeed so that ground effect can be maintained.

3. Throughout the maneuver, maintain a constant groundspeed and ground track with cyclic stick, a constant heading with pedals, and a constant altitude with collective pitch control .along with throttle coordination to maintain proper RPM.

4. To stop rearward movement, apply forward cyclic pressure and hold until the helicopter stops. As motion stops, return the cyclic stick to the neutral position. Also, as in the case of forward and sideward flight, forward cyclic can be used to level the helicopter and let it drift to a stop.

COMMON ERRORS:

1. Erratic movement of the cyclic stick, resulting in overcontrol and an uneven movement over the ground.

2. Failure to use pedals properly, resulting in excessive nose movements.

3. Failure to maintain hovering altitude.

4. Failure to maintain proper RPM.

5. Failure to make clearing turns prior to starting the maneuver.

TAXIING

Taxiing is the intentional movement of the helicopter, under its own power, while remaining in contact with the surface.

TECHNIQUE:

1. The helicopter should be in a stationary position on the surface with the collective pitch full down, and the RPM the same as that which is used for hover operations.

2. Move the cyclic slightly forward of the neutral position and apply a gradual upward pressure on the collective pitch to move forward along the surface. Use pedals to maintain heading and cyclic to maintain ground track.

3. The collective pitch controls starting, stopping, and rate of speed while taxiing. The higher the collective pitch, the faster the taxi speed. Never taxi at a speed greater than a normal walk.

4. During crosswind taxi, the cyclic should be held into the wind a sufficient amount to eliminate any drifting movement.

5. Maintain proper RPM at all times.

COMMON ERRORS:

1. Improper use of cyclic stick—using the cyclic to control starting, stopping, and rate of speed.

2. Failure to use pedals for heading control.

3. Improper use of controls during crosswind operations.

4. Failure to maintain proper RPM.

NORMAL TAKEOFF FROM A HOVER

Takeoff from a hover is an orderly transition to forward flight and is executed to increase altitude safely and expeditiously.

TECHNIQUE:

1. Lift the helicopter to a hover. Check engine and control operation. Note the cyclic stick position to determine if the aircraft is loaded properly. Check the manifold pressure required to hover to determine the amount of excess power available.

2. Clear the area all around.

3. Smoothly and slowly ease the cyclic stick forward. Apply just enough forward cyclic pressure to start the helicopter moving forward over the ground (fig. 70).

4. As the helicopter starts to move forward, increase collective pitch as necessary to prevent settling upon departing from ground effect and adjust throttle to maintain RPM. The increase in power will require an increase in left pedal pressure to maintain heading. A straight takeoff path should be maintained throughout the takeoff.

5. As you accelerate to effective translational lift, the helicopter will begin to climb and the nose will tend to rise due to increased lift. At this point adjust collective pitch to obtain normal climb power and apply enough forward cyclic stick to overcome the tendency of the nose to rise. Hold an attitude that will allow a smooth accelera-

● 1. HOVER AT NORMAL HOVERING ALTITUDE.   ● 2. 3. & 4. EASE CYCLIC FORWARD, INCREASE COLLECTIVE PITCH TO PREVENT SETTLING.   ● 5. 6. & 7. AS AIRSPEED APPROACHES NORMAL CLIMB SPEED, RAISE NOSE TO CLIMBING ATTITUDE.   ● 8. & 9. CONTINUE CLIMB AT NORMAL CLIMB SPEED.

FIGURE 70.—Normal takeoff from a hover.

tion toward climbing airspeed and a commensurate gain in altitude so that the takeoff profile will not take you through any of the cross-hatched area of the height–velocity chart for the particular helicopter. As airspeed is increased, the streamlining of the fuselage reduces engine torque effect, requiring a gradual reduction of left pedal.

6. As the helicopter continues to climb and airspeed approaches normal climb speed, apply aft cyclic stick pressure to raise the nose smoothly to the normal climb attitude.

COMMON ERRORS:

1. Failing to use sufficient collective pitch to prevent settling between the time the helicopter leaves ground effect and picks up translational lift.

2. Adding power too rapidly at the beginning of the transition from hovering to forward flight without forward cyclic compensation, causing the helicopter to gain excessive altitude before acquiring airspeed.

3. Assuming an extreme nose-down attitude near the ground in the transition from hovering to forward flight.

4. Failing to maintain a straight flight path over the ground (ground track).

5. Failing to maintain proper airspeed during the climb.

6. Failing to adjust the throttle to maintain proper RPM.

Crosswind considerations during takeoffs

If a takeoff is made in a crosswind, the helicopter is flown in a slip (p. 79) during the early stages of the maneuver. The cyclic pitch is held into the wind to maintain the selected ground track for the takeoff, and the heading is kept straight along the takeoff path with the antitorque pedals (fig. 71). Thus, the ground track and fuselage are alined with each other. In other words, the rotor is tilted into the wind so that the sideward movement of the helicopter is just enough to counteract the wind drift. To prevent the nose from turning in the direction of rotor tilt, it will be necessary to increase pedal pressure on the side opposite to rotor tilt. The stronger the crosswind component, the greater the amount of opposite pedal pressure required to maintain heading.

After approximately 50 feet of altitude is gained, a heading (crab) into the wind (fig. 71) should be established by a coordinated turn to maintain the desired ground track. The stronger the crosswind component, the more the helicopter will have to be turned into the wind to maintain desired ground track. Once straight-and-level flight on the desired heading is obtained, the pedals should continue to be used as necessary to compensate for torque to keep the helicopter in longitudinal trim.

NORMAL TAKEOFF FROM THE SURFACE

The normal takeoff from the surface is used to move the helicopter from a position on the surface into effective

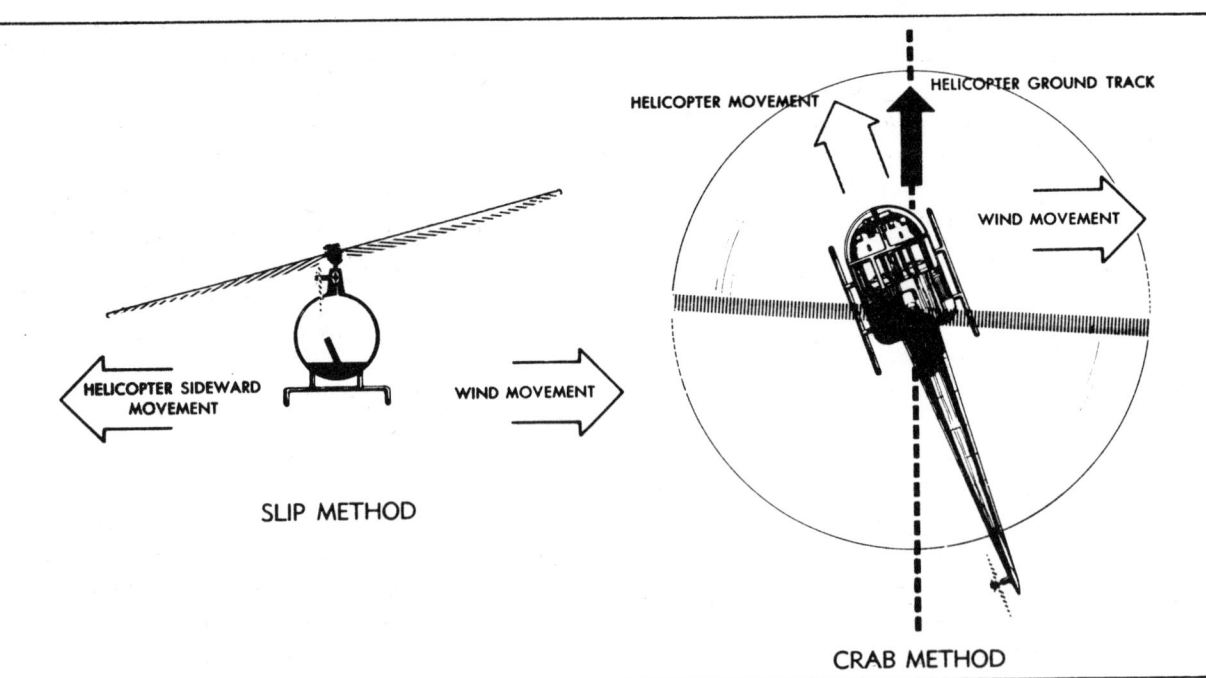

SLIP METHOD

CRAB METHOD

Figure 71.—Comparison of the slip method and crab method of wind drift correction.

translational lift and a normal climb using a minimum amount of power.

TECHNIQUE:

1. Place the helicopter in a stationary position on the surface, lower the collective pitch to the full down position and reduce the RPM below operating RPM. Visually clear the area and select terrain features or other objects to aid in maintaining the desired track during the takeoff and climbout.

2. Increase the throttle to proper RPM and raise the collective pitch slowly until the helicopter is light on the skids. Hesitate momentarily and adjust cyclic and pedals as necessary to prevent any surface movement.

3. Continue to apply upward collective pitch and as the helicopter breaks ground, use cyclic as necessary to assure forward movement as altitude is gained.

4. Continue to accelerate the aircraft and, as effective translational lift is attained, the helicopter will begin to climb. Adjust attitude and power, if necessary, to climb in the same manner as a takeoff from a hover.

COMMON ERRORS:

1. Departing the surface in an attitude that is too nose-low. This situation requires the use of excessive power to initiate a climb.

2. Using excessive power combined with too level an attitude which causes a vertical climb.

3. Too abrupt application of the collective pitch when departing the surface, causing RPM and heading control errors.

STRAIGHT-AND-LEVEL FLIGHT

Straight-and-level flight is flight in which a constant altitude and heading are maintained. (The straight-and-level flight attitude is the attitude of the helicopter necessary to maintain straight-and-level flight. The level-flight attitude is the attitude of the helicopter necessary to maintain altitude. These terms are used throughout this handbook.)

The airspeed is determined by the attitude of the helicopter. The attitude of the helicopter is controlled by the movement of the cyclic control stick; altitude is primarily maintained by use of the collective pitch. In order to maintain forward flight, the rotor tip-path plane must be tilted forward to obtain the necessary horizontal thrust component from the main rotor. This will generally result in a nose-low attitude. The lower the nose, the greater the power required to maintain altitude (and the higher the resulting speed). Conversely, the greater the power used, the lower the nose must be to maintain altitude.

When in straight-and-level flight, an increase in collective pitch while holding airspeed constant with the cyclic control causes the helicopter to climb; a decrease in collective pitch while holding airspeed constant causes a descent. A correction on the collective pitch control requires coordinated correction on the throttle control in order to maintain a constant RPM and on the antitorque pedals to maintain heading and to keep the helicopter in longitudinal trim. Coordinated flight should be maintained (that is, neither slipping nor skidding).

To increase airspeed in straight-and-level flight, gently apply forward pressure on the cyclic control stick and raise the collective pitch as necessary to maintain altitude. To decrease airspeed, gently apply aft pressure on the cyclic control stick and lower the collective pitch as necessary to maintain altitude.

Although cyclic pitch control is sensitive, there is a slight delay in control reaction and it will be necessary to anticipate actual movement of the helicopter. In making cyclic corrections to control the attitude or airspeed of a helicopter, care should be taken not to overcontrol. If the nose of the helicopter rises above the level-flight attitude, forward pressure is applied to the cyclic stick to bring the nose down. If this correction is held too long the nose will drop too low. Since the helicopter will continue to change attitude momentarily after the controls reach neutral, return the cyclic stick control to neutral slightly before the desired attitude is reached. This principle holds true for any cyclic pitch control correction.

The helicopter is inherently unstable. If gusts or turbulence cause the nose to drop, it will tend to continue dropping instead of returning to a straight-and-level attitude as would a fixed-wing aircraft. A pilot must remain alert and FLY the helicopter at all times.

COMMON ERRORS:

1. Failure to trim properly, tending to hold pedal pressure and opposite cyclic.

2. Failure to hold best airspeed. Aft cyclic pressure dissipates airspeed without significant climb.

3. Failure to recognize proper control position for maintaining crab-type drift correction.

TURNS

A turn is a maneuver used to change the heading of the helicopter. The aerodynamics of a turn have been discussed previously—lift components, loss of vertical lift, and load factors—and should be thoroughly understood.

Before beginning any turn, the area in the direction of the turn should be carefully cleared above, below, and at the flight level. To enter a turn from straight-and-level flight, apply sideward pressure on the cyclic stick in the direction the turn is to be made. This is the only control movement necessary to start the turn. Do not use the pedals to assist the turn. The pedals should be used to compensate for torque to keep the helicopter in longitudinal trim. The more the cyclic stick is displaced, the steeper the angle of bank; therefore, adjust the cyclic stick to obtain and maintain the desired bank throughout the turn. Increase collective pitch as necessary to maintain altitude at the same time coordinating throttle to maintain RPM and increase left pedal pressure to counteract the added torque effect and to maintain longitudinal trim. Depending on the degree of bank, additional forward cyclic pressure may be required to maintain airspeed.

Recovery from the turn is the same as the entry except that pressure on the cyclic stick is applied in the opposite direction. Since the helicopter will continue to turn as long as there is any bank, start the rollout before reaching the desired heading.

The discussion on level turns is equally applicable to making turns while climbing or descending, the only difference being that the helicopter will be in a climbing or descending attitude rather than the level flight attitude. If a simultaneous entry is desired, merely combine the techniques of both maneuvers—climb or descent entry and turn entry.

Skids

A skid occurs when the helicopter slides sideways away from the center of the turn. It is caused by too much pedal pressure in the direction of turn or too little in the direction opposite the turn, in relation to the amount of collective stick (power) used. If the helicopter is forced to turn faster with increased pedal pressure instead of increasing the degree of bank, it will skid sideways away from the center of turn. Instead of flying in its normal curved pattern, it will fly a straighter course.

In a right climbing turn, if insufficient left pedal is applied to compensate for increased torque effect, a skid will occur. In a left climbing turn, if excessive left pedal is applied to compensate for increased torque effect, a skid will occur.

In a right descending turn, if excessive right pedal is applied to compensate for decreased torque, a skid will occur. In a left descending turn, if insufficient right pedal is applied to compensate for the decreased torque effect, a skid will occur.

A skid may also occur when flying straight-and-level if the nose of the helicopter is allowed to move sideways along the horizon. This condition occurs when improper pedal pressure is held to counteract torque and the helicopter is held level with cyclic control.

Slips

A slip occurs when the helicopter slides sideways toward the center of the turn. It is caused by an insufficient amount of pedal in the direction of turn (or too much in the direction opposite the turn) in relation to the amount of collective stick (power) used. In other words, if improper pedal pressure is held, keeping the nose from following the turn, the helicopter will slip sideways toward the center of turn.

In a right climbing turn, if excessive left pedal is applied to compensate for the increased torque effect, a slip will occur. In a left climbing turn, if insufficient left pedal is applied to compensate for the increased torque effect, a slip will occur.

In a right descending turn, if insufficient right pedal is applied to compensate for the decreased torque effect, a slip will occur. In a left descending turn, if excessive right pedal is applied to compensate for the decreased torque effect, a slip will occur.

A slip may also occur in straight-and-level flight if one side of the helicopter is low and the nose is held straight by pedal pressure. This is the technique used in correcting for a crosswind during an approach and during a takeoff when at a low altitude.

Summarizing then, a skid occurs when the rate of turn is too fast for the amount of bank; a slip occurs when the rate of turn is too slow for the amount of bank.

COMMON ERRORS:

1. Failure to hold altitude.

2. Using pedal pressures for turns. This is not necessary for small helicopters.

NORMAL CLIMB

The entry into a climb from a hover has already been discussed under "Normal Takeoff From a Hover." This discussion will be limited to a climb entry from cruising flight.

TECHNIQUE:

1. To enter a climb from cruising flight, apply aft pressure on the cyclic stick to obtain the approximate climb attitude; simultaneously increase collective pitch to obtain climb manifold pressure, adjust throttle to maintain or obtain climb RPM, and increase left pedal pressure to compensate for the increased torque.

2. As the airspeed approaches normal climb airspeed, make further adjustments of the cyclic control to obtain and hold this airspeed.

3. Throughout the maneuver, maintain climb attitude and airspeed with cyclic control, climb manifold pressure

and RPM with collective pitch and throttle, and longitudinal trim and heading with antitorque pedals.

4. To level off from a climb, start adjusting attitude to the level flight attitude a few feet prior to reaching the desired altitude. The amount of lead will depend upon the rate of climb at the time of level-off—the higher the rate of climb, the more the lead. Apply forward cyclic pressure to adjust to and maintain the level flight attitude which will be slightly nose low; maintain climb power until airspeed approaches desired cruising airspeed at which time the collective should be lowered to obtain cruising manifold pressure and throttle adjusted to obtain and maintain cruising RPM. Throughout the level-off, maintain longitudinal trim and a constant heading with pedals.

COMMON ERRORS:

1. Failure to hold proper manifold pressure and airspeed.

2. Holding too much or too little left pedal pressure.

3. In level-off, decreasing power before lowering nose to cruising attitude.

NORMAL DESCENT

A normal descent is a maneuver in which the helicopter loses altitude at a controlled rate and while in a controlled attitude.

TECHNIQUE:

1. To establish a normal descent from straight-and-level flight at cruising airspeed, lower collective pitch to obtain proper manifold pressure, adjust throttle to maintain RPM, and increase right pedal pressure to maintain heading. If cruising airspeed is the same as, or slightly above descending airspeed, simultaneously apply the necessary cyclic stick pressure to obtain the approximate descending attitude. If cruising airspeed is well above descending airspeed, the level flight attitude may be maintained until airspeed approaches descending airspeed, at which time the nose should be lowered to the descending attitude.

2. Throughout the maneuver, maintain descending attitude and airspeed with the cyclic control, descending manifold pressure and RPM with collective pitch and throttle, and heading with pedals.

3. To level off from the descent, lead the desired altitude by an amount that will depend upon the rate of descent at the time of level-off, for example, the higher the rate of descent, the greater the lead. At this point, increase collective pitch to obtain cruising manifold pressure, adjust throttle to maintain RPM, increase left pedal pres-

sure to maintain heading, and adjust cyclic stick to obtain cruising airspeed and the level-flight attitude as the desired altitude is reached.

COMMON ERRORS:

1. Failure to hold constant angle of descent (training purposes only).

2. Failure to adjust pedal pressures for changes in power.

APPROACHES

An approach is a transition maneuver in which the helicopter is flown from traffic pattern altitude at cruising airspeed to a hover at normal hovering altitude and with zero groundspeed. It is basically a power glide made at an angle of descent corresponding to the type of approach that is made.

A helicopter pilot should be proficient in performing three basic types of approaches—normal, steep, and shallow. He should know how to analyze influential outside factors and to know how to plan an approach to fit any particular situation. Choice of approach is governed by the size of the landing area, barriers in the approach path, type of ground surface, temperature, altitude, density altitude, wind direction, wind speed, and gross weight.

All approaches should be regarded as precision approaches and should be made to a predetermined point. Rate of descent and airspeed are independently controlled by the pilot. Therefore, little tolerance should be given to overshooting or undershooting a chosen landing spot. In order to maintain a maximum margin of safety in each type of approach, effective translational lift should be retained as long as practicable.

Factors to consider when making approaches

Evaluation of existing wind conditions must be made before initiating an approach. Although the approach is generally made into the wind, conditions may indicate that entry will have to be made from a downwind or crosswind position. The traffic pattern is generally flown at normal cruise airspeed. The velocity of the wind determines the airspeed that will be maintained after the approach is initiated. Airspeed should be increased in proportion to any increased wind velocity. Angle of descent should remain constant regardless of wind velocity.

Before attempting normal and steep approaches to a hover, the pilot should know that sufficient hovering power is available. For a shallow approach terminating in a running landing, a ground area of sufficient length and smoothness must be available.

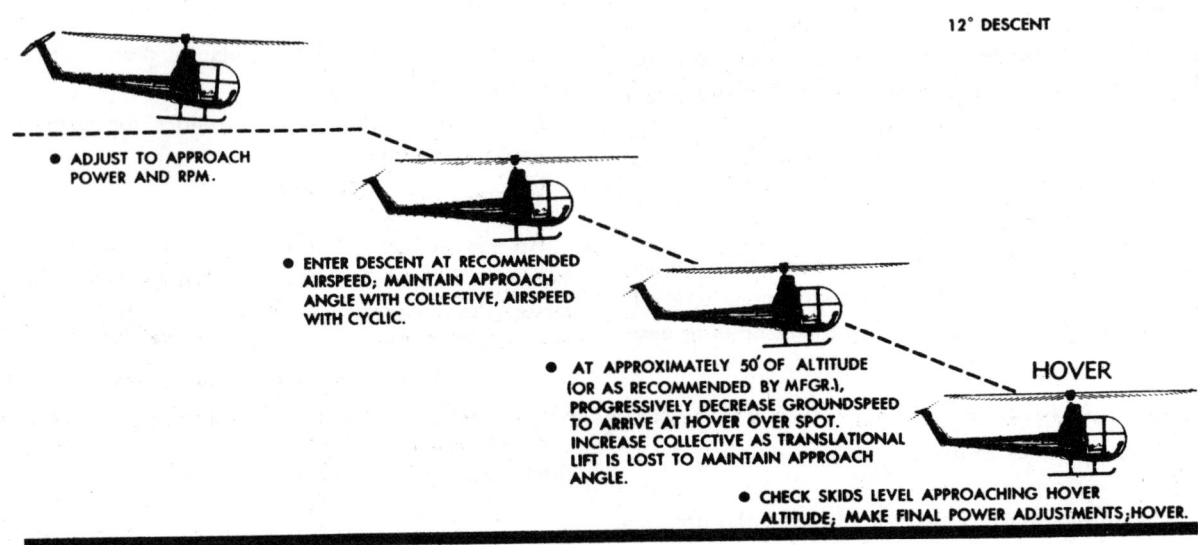

FIGURE 72.—Normal approach to a hover.

Crosswind approaches are made by crabbing or slipping or a combination of both. To make running landings in strong crosswinds, it may be necessary to touch down initially with the windward (upwind) skid to avoid drifting.

RPM should remain constant during all approaches. If RPM is allowed to fluctuate or change abruptly, variations of torque forces will cause the fuselage to yaw around the vertical axis and control will be difficult. To maintain proper directional control, changes in RPM and/or collective pitch settings must be made smoothly and must be accompanied by appropriate changes in antitorque pedals.

NORMAL APPROACH TO A HOVER

A normal approach to a hover is basically a power glide made at an angle of descent of approximately 12° (fig. 72). This type of approach is used in the majority of cases.

TECHNIQUE:

1. The entry to the downwind leg should be made at a 45° angle to the downwind leg so that the actual turn to the downwind leg will be accomplished opposite the middle one-third of the runway. The transition from downwind leg to the final approach leg may be made by two 90° turns in which a definite base leg is established, or by a 180° turn. At all times during this transition, sufficient altitude should be available so that in case of engine failure, an autorotative landing can be completed into the wind. The point in the traffic pattern at which the power reduction is made should be determined by this fact.

2. Initiate the approach by lowering the collective pitch control the amount required to descend at an angle of approximately 12° on the final approach leg. As collective pitch is lowered, increase right pedal pressure as necessary to compensate for the change in torque reaction to maintain heading and adjust throttle to maintain proper RPM. Decelerate to the approximate airspeed, then further adjust attitude as necessary to maintain approach airspeed.

3. The angle of descent is primarily controlled by collective pitch, the airspeed is primarily controlled by the cyclic control, and heading on final approach is maintained with pedal control. However, only by the coordination of all controls can the approach be accomplished successfully.

4. The approach airspeed should be maintained until the point on the approach is reached where, through

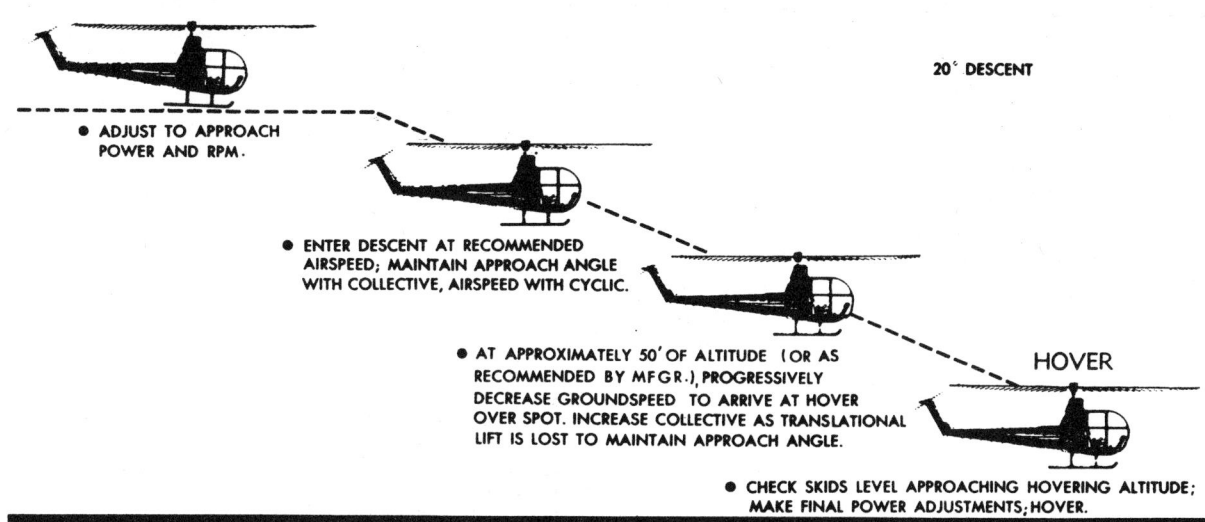

20° DESCENT

● ADJUST TO APPROACH POWER AND RPM.

● ENTER DESCENT AT RECOMMENDED AIRSPEED; MAINTAIN APPROACH ANGLE WITH COLLECTIVE, AIRSPEED WITH CYCLIC.

● AT APPROXIMATELY 50' OF ALTITUDE (OR AS RECOMMENDED BY M.F.G.R.), PROGRESSIVELY DECREASE GROUNDSPEED TO ARRIVE AT HOVER OVER SPOT. INCREASE COLLECTIVE AS TRANSLATIONAL LIFT IS LOST TO MAINTAIN APPROACH ANGLE.

HOVER

● CHECK SKIDS LEVEL APPROACHING HOVERING ALTITUDE; MAKE FINAL POWER ADJUSTMENTS; HOVER.

FIGURE 73.—Steep approach to a hover.

evaluation of apparent groundspeed, it is determined that forward airspeed must be progressively decreased in order to arrive at hovering altitude and attitude at the intended landing spot with zero groundspeed.

5. As forward airspeed is gradually reduced by the application of rearward cyclic pressure, additional power (collective pitch) must be applied to compensate for the decrease in translational lift and to maintain the proper angle of descent. As collective pitch is increased, left pedal pressure must be increased to maintain heading, throttle adjusted to maintain RPM, and cyclic pitch co-ordinated to maintain the proper rate of closure to the desired spot (a continual decrease in groundspeed).

6. The approach is terminated at hovering altitude above the intended landing point with zero ground-speed. If power has been properly applied during the final portion of the approach, very little additional power should be required during the termination.

7. If the condition of the landing spot is unknown, the approach may be terminated just short of the spot so that it can be checked before moving forward for the landing.

COMMON ERRORS:

1. Failing to maintain proper RPM during the entire approach.

2. Improper use of the collective pitch in controlling the angle of descent.

3. Failing to make pedal corrections to compensate for collective pitch changes during the approach.

4. Failing to arrive at hovering altitude, hovering attitude, and zero groundspeed almost simultaneously.

5. Low RPM in transition to the hover at the end of the approach.

6. Using too much aft cyclic stick close to the ground which may result in the tail rotor striking the ground.

Crosswind considerations in approaches

During the early stages of a crosswind approach, a crab and/or a slip may be used (fig. 71). During the final stages of an approach, beginning at approximately 50 feet of altitude, a slip should be used to aline the fuselage with the ground track. The rotor is tilted into the wind (with cyclic pressure) enough so that the sideward movement of the helicopter and wind drift counteract each other. Heading is maintained along the ground track with the antitorque pedals. (See "Crosswind Considerations During Takeoffs.") This technique should be used on any type of crosswind approach—shallow, normal, or steep.

NORMAL APPROACH TO THE SURFACE

When it is known or suspected that loose snow or dust exists on your landing spot, an approach to the surface may be used. It may also be used when the surface is unfavorable for a running landing, and high density altitude or heavily loaded conditions exist.

TECHNIQUE:

The approach is the same as the normal approach to a hover. However, the approach should be continued to touchdown, terminating in a skids-level attitude with no forward movement.

COMMON ERRORS:

1. Terminating at a hover, then making a vertical landing.

2. Touching down with forward movement.

3. Approaching too slow, requiring the use of excessive power during the termination.

4. Approaching too fast, causing a hard landing.

STEEP APPROACH TO A HOVER

A steep approach is used primarily when there are obstacles in the approach path that are too high to allow a normal approach. A steep approach will permit entry into most confined areas and is sometimes used to avoid areas of turbulence around a pinnacle. An approach angle of between 12° and 20° is normally used for steep approaches (fig. 73).

TECHNIQUE:

1. Entry is made in the same way as for a normal approach, except that a greater reduction of collective pitch is usually required at the beginning of the approach to start the descent than for a normal approach. As collective pitch is lowered, increase right pedal pressure to maintain heading and adjust throttle to maintain RPM.

2. As in a normal approach, the angle of descent is primarily controlled by collective pitch, and the speed is primarily controlled by the cyclic control. However, only by the coordination of all controls can the approach be accomplished successfully.

3. The approach airspeed should be maintained until the point on the approach is reached where, through evaluation of apparent groundspeed, it is determined that forward airspeed must be progressively decreased in order to arrive at hovering altitude at the intended landing spot with zero groundspeed. This is very important since a flare should not be made near the ground due to the danger of the tail rotor striking the ground.

4. As forward airspeed is gradually reduced by the application of rearward cyclic pressure, additional power (collective pitch) must be applied to compensate for the decrease in translational lift and to maintain the proper angle of descent. As collective pitch is increased, left pedal pressure must be increased to maintain heading, throttle adjusted to maintain RPM, and cyclic pitch coordinated to maintain the proper change in forward airspeed.

5. Since the angle of descent on a steep approach is much steeper than for a normal approach, the collective pitch must be used much sooner at the bottom of the approach. The approach is terminated at hovering altitude above the intended landing point with zero groundspeed. If power has been properly applied during the final portion of the approach, very little additional power should be required during the termination.

COMMON ERRORS:

1. Failing to maintain proper RPM during the entire approach.

2. Improper use of collective pitch in controlling the angle of descent.

3. Failing to make pedal corrections to compensate for collective pitch changes during the approach.

4. Slowing airspeed excessively in order to remain on the proper angle of descent.

5. Failing to arrive at hovering altitude, hovering attitude, and zero groundspeed almost simultaneously.

6. Low RPM in transition to the hover at the end of the approach.

7. Using too much aft cyclic stick close to the ground which may result in the tail rotor striking the ground.

LANDING FROM A HOVER

In this maneuver, the helicopter is landed vertically from a hover.

TECHNIQUE:

1. From a hover, begin a descent by applying a slow but very gradual downward pressure on the collective pitch stick. This smooth application of collective pitch should be such that a constant rate of descent is maintained to the ground. As the skids descend to within a few inches of the ground, the effect of ground cushion becomes very noticeable and the helicopter tends to stop its descent. At this point, it may be necessary to further decrease the collective pitch stick a slight amount in order to maintain the constant rate of descent.

2. When the skids touch the ground, lower the collective pitch smoothly and firmly to the full down position, adjust the throttle to keep RPM in the proper range, and at the same time add right pedal pressure as needed to maintain heading.

3. Throughout the descent and until the time the skids are firmly on the ground and the collective pitch is in full down position, make necessary corrections with pedals to maintain a constant heading, and necessary corrections with the cyclic control to maintain a level attitude and prevent movement over the ground.

COMMON ERRORS:

1. Overcontrolling the cyclic control during descent resulting in movement over the ground on contact.

2. Failing to use collective pitch smoothly.

3. Pulling back on the cyclic stick prior to or upon touchdown.

4. Failing to reduce the collective pitch smoothly and positively to the full down position upon contact with the ground.

5. Failing to maintain a constant rate of descent.

6. Failing to maintain proper RPM.

SHALLOW APPROACH AND RUNNING LANDING

A shallow approach and running landing (fig. 74) are used when a high density altitude or a high gross weight condition or some combination thereof is such that a normal or steep approach cannot be made because of insufficient power to hover. To compensate for this lack of power, a shallow approach and running landing makes use of translational lift until ground contact is made. The glide angle is from 5° to 12°, depending on wind conditions. Since a running landing follows a shallow approach, a ground area of sufficient length and smoothness must be available.

TECHNIQUE:

1. A shallow approach is initiated in the same manner as the normal approach except that a shallower angle of descent is maintained. The power reduction to initiate the desired angle of descent will be less than that for a normal approach since the angle of descent is less. As collective pitch is lowered, maintain heading by increasing right pedal pressure, adjust throttle to maintain RPM, and use cyclic as necessary to maintain the desired approach airspeed.

2. As in normal and steep approaches, the angle of descent and rate of descent are primarily controlled by collective pitch, and the groundspeed is primarily controlled by the cyclic control. The coordination of all controls is needed, however, if the approach is to be accomplished successfully.

3. Approach airspeed should be maintained until an altitude of approximately 50 feet above the ground has been reached. At this point, very gradually apply aft cyclic stick to start dissipating airspeed and coordinate a slight downward pressure on the collective pitch to maintain the angle of descent. The deceleration of the airspeed should be enough so that the helicopter will tend to descend to the ground due to the decreased effect of translational lift just as the landing spot is reached. Since translational lift diminishes rapidly at slow airspeeds, the deceleration must be smoothly coordinated, at the same time keeping enough lift to prevent the helicopter from settling abruptly.

4. On the final part of the approach, prior to making ground contact, the helicopter should be placed in a level attitude with cyclic control, pedals should be used to maintain heading, and cyclic stick should be used as necessary so that heading and ground track are identical. Allow the helicopter to descend gently to the ground in a

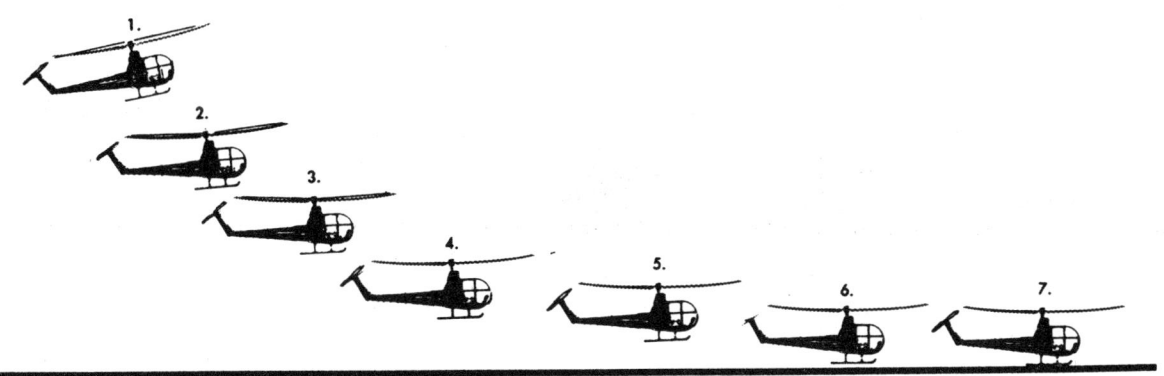

● 1. APPROXIMATELY 50 FT. RAISE NOSE SLIGHTLY ABOVE LEVEL ATTITUDE, DISSIPATE AIRSPEED. ● 2. & 3. LEVEL HELICOPTER AND USE COLLECTIVE AS NECESSARY TO SLOW RATE OF DESCENT. ● 4. 5. & 6. USE COLLECTIVE TO CUSHION LANDING. ● 7. CONTINUE SKIDDING.

FIGURE 74.—Running landing.

straight-and-level attitude, cushioning the landing by proper manipulation of the collective pitch.

5. After ground contact, the cyclic control should be placed slightly forward of neutral to tilt the main rotor away from the tail boom; antitorque pedals should be used to maintain heading; throttle should be used to maintain RPM; and cyclic stick should be used to maintain ground track. Normally, the collective pitch is held stationary after touchdown until the helicopter comes to a complete stop. However, if braking action is desired or required, the collective pitch may be lowered cautiously. To ensure directional control, normal rotor RPM must be maintained until the helicopter stops.

COMMON ERRORS:

1. Assuming excessive nose-high attitude at approximately 10 feet.

2. Insufficient collective pitch and throttle to cushion landing.

3. Failing to add left pedal as collective pitch is added to cushion landing resulting in a touchdown while in a left skid.

4. Touching down at an excessive groundspeed for the existing conditions (20 MPH groundspeed in most cases would be considered maximum allowable).

5. Failing to touch down in a level attitude.

6. Failing to maintain proper rotor RPM during and after touchdown.

7. Poor directional control upon touchdown.

RUNNING TAKEOFF

A running takeoff (fig. 75) is used when conditions of load and/or density altitude prevent a sustained hover at normal hovering altitude. It is often referred to as a high-altitude takeoff. With insufficient power to hover, at least momentarily or at a very low altitude, a running takeoff is not advisable. No takeoff should be attempted if the helicopter cannot be lifted off the ground momentarily at full power because:

1—If the helicopter will not hover, its performance is unpredictable.

2—If the helicopter cannot be raised off the ground at all, sufficient power might not be available for a safe running takeoff.

A running takeoff may be accomplished safely only if ground area of sufficient length and smoothness is available, and if no barriers exist in the flight path to interfere with a shallow climb.

TECHNIQUE:

1. Head the helicopter into the wind.

2. Increase the throttle to obtain takeoff RPM.

3. Hold cyclic stick slightly forward of the hovering neutral position. Apply collective pitch slowly to accelerate into forward movement. (For practice a manifold pressure of 1 to 2 inches below that required to hover may be used.)

4. Maintain a straight ground track with lateral cyclic control and heading with antitorque pedals until a climb is established.

● 1. & 2. ADJUST POWER (1"-2" LESS THAN HOVERING POWER); CYCLIC SLIGHTLY FORWARD OF HOVERING POSITION FOR GRADUAL ACCELERATION.

● 3. AT TRANSLATIONAL LIFT, EASE CYCLIC BACK SLIGHTLY TO BECOME AIRBORNE.

● 4. MAINTAIN ALTITUDE AT 10' OR LESS TO BUILD TO CLIMB AIRSPEED.

● 5. ADJUST TO CLIMB ALTITUDE.

FIGURE 75.—High altitude (running) takeoff.

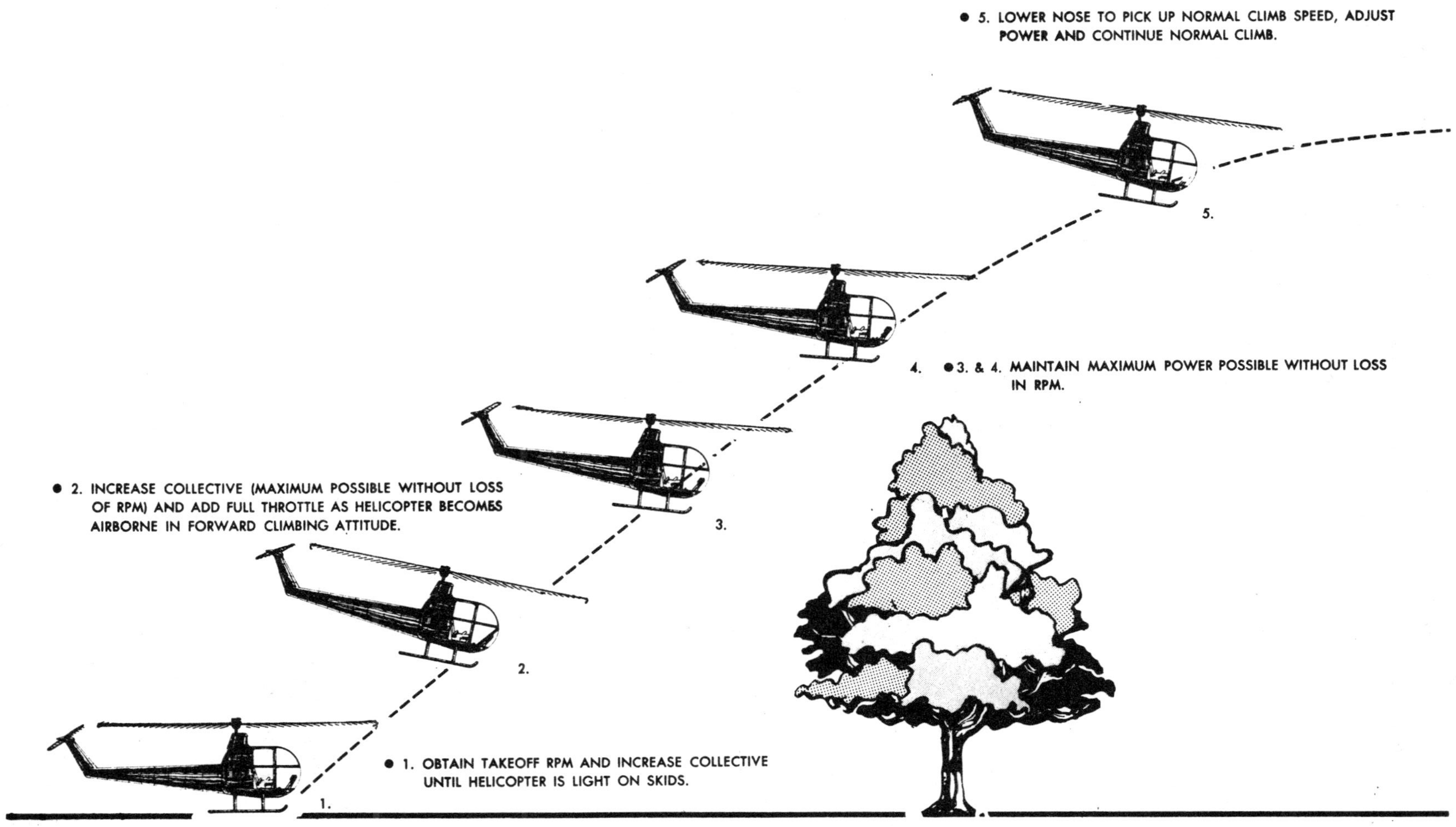

5. LOWER NOSE TO PICK UP NORMAL CLIMB SPEED, ADJUST POWER AND CONTINUE NORMAL CLIMB.

3. & 4. MAINTAIN MAXIMUM POWER POSSIBLE WITHOUT LOSS IN RPM.

2. INCREASE COLLECTIVE (MAXIMUM POSSIBLE WITHOUT LOSS OF RPM) AND ADD FULL THROTTLE AS HELICOPTER BECOMES AIRBORNE IN FORWARD CLIMBING ATTITUDE.

1. OBTAIN TAKEOFF RPM AND INCREASE COLLECTIVE UNTIL HELICOPTER IS LIGHT ON SKIDS.

Figure 76.—Maximum performance takeoff.

5. As effective translational lift is gained, slight back pressure on the cyclic stick will take the helicopter into airborne flight smoothly, in a level attitude, with little or no pitching.

6. Maintain an altitude not to exceed 10 feet to allow airspeed to increase toward normal climb speed and follow a climb profile that will take you through the clear area of the height–velocity curve for the particular helicopter.

7. During practice maneuvers, climb to 50 feet then adjust power to normal climb power and attitude to normal climb attitude.

COMMON ERRORS:

1. Failing to aline heading and ground track to keep ground friction to a minimum.

2. Attempting to pull the helicopter off the ground before effective translational lift is obtained.

3. Lowering the nose too much after becoming airborne resulting in the helicopter settling back to the ground again.

4. Failing to remain below approximately 10 feet of altitude until airspeed approaches normal climb speed.

MAXIMUM PERFORMANCE TAKEOFF

A maximum performance takeoff is used to climb at a steep angle in order to clear barriers in the flight path (fig. 76). It can be used when taking off from small fields surrounded by high obstacles. Before attempting a maximum performance takeoff, you must know thoroughly the capabilities and limitations of your equipment. You must take into consideration the wind velocity, temperature, altitude, density altitude, gross weight, center-of-gravity location, and other factors affecting your technique and the performance of the helicopter. To safely accomplish this type of takeoff, sufficient power to hover must be available to prevent the helicopter from sinking back to the ground after becoming airborne. This maneuver will result in a steep climb, affording maximum altitude gain in a minimum distance forward.

The angle of climb for a maximum performance takeoff will depend on existing conditions. The more critical the conditions—high density altitudes, calm winds, etc.—the shallower the angle of climb should be. Use caution in climbing steeply. If the airspeed is allowed to get too low, the helicopter may settle back to the ground. The height–velocity (H/V) chart for the particular helicopter should be fully considered before making any maximum performance takeoff. An engine failure at low altitude and airspeed would place the helicopter in a dangerous position, requiring a high degree of skill in making a safe autorotative landing. It may be necessary to operate in the shaded area of the H/V diagram during the beginning of this maneuver when operating in light or no-wind conditions. The angle of climb and resulting airspeed will be dictated by the proximity and height of the obstacles to be cleared. The pilot must be aware of the calculated risk involved when operating in the shaded area of the H/V diagram.

TECHNIQUE:

1. The helicopter should be headed generally into the wind and the cyclic stick placed in what would be the neutral position for hovering under the existing load and wind conditions. (This position could be checked by hovering the helicopter momentarily prior to preparing to execute a maximum performance takeoff.)

2. Establish the proper RPM setting and apply sufficient collective pitch to lighten the helicopter on its landing gear (fig. 76). Apply the maximum amount of collective pitch that can be obtained without reducing RPM and simultaneously add full throttle and sufficient forward cyclic stick to establish a forward climbing attitude as the helicopter leaves the ground. Apply necessary antitorque pedal control to maintain heading. RPM must not be sacrificed in order to obtain increased pitch on the rotor blades. If RPM starts to decrease under a full power condition, it can be regained only by reducing collective pitch.

3. Utilize full power until the helicopter is clear of all obstacles, after which a normal climb can be established and power reduced.

COMMON ERRORS:

1. Too much forward cyclic stick initially allowing the nose to go down too far.

2. Failure to maintain maximum permissible RPM.

3. Control movements too abrupt.

AUTOROTATIONS

In helicopter flying, an autorotation is a maneuver that can be performed by the pilot whenever the engine is no longer supplying power to the main rotor blades. A helicopter transmission is designed to allow the main rotor to rotate freely in its original direction if the engine stops. At the instant of engine failure, the blades will be producing lift and thrust from their angle of attack and velocity. By immediately lowering collective pitch (which must be done in case of engine failure), lift and drag will be reduced, and the helicopter will begin immediate descent, thus producing an upward flow of air through the rotor system. The impact of this upward flow of air on the rotor blades produces a "ram" effect which provides sufficient thrust to maintain rotor RPM throughout the descent. Since the tail rotor is driven by the main rotor during autorotation, heading control can be maintained as in normal flight.

Several factors affect the rate of descent in autorotation—air density (density altitude), gross weight, rotor RPM, and airspeed. The pilot's primary control of the rate of descent is the airspeed. Higher or lower airspeed is obtained with the cyclic control just as in normal flight. A pilot has a choice in angle of descent varying from vertical descent to maximum range (minimum angle of descent). Rate of descent is high at zero airspeed and decreases to a minimum somewhere in the neighborhood of 50 to 60 miles per hour depending upon the particular helicopter and the factors just mentioned. As the airspeed increases beyond that which gives minimum rate of descent, the rate of descent increases again. When an autorotative landing is to be made, the energy stored in the rotating blades can be used by the pilot to decrease the rate of descent and make a safe landing. A greater amount of rotor energy is required to stop a helicopter with a high rate of descent than is required to stop a helicopter that is descending more slowly. It follows then that autorotative descents at very low or very high airspeeds are more critical than those performed at the proper airspeed for the minimum rate of descent.

Each type of helicopter has a specific airspeed at which a power-off glide is most efficient. The best airspeed is the one which combines the most desirable (greatest) glide range with the most desirable (slowest) rate of descent. The specific airspeed is somewhat different for each type of helicopter, yet certain factors affect all configurations in the same manner. For specific autorotation airspeeds for a particular helicopter refer to the helicopter flight manual.

The specific airspeed for autorotations is established for each type of helicopter on the basis of average weather and wind conditions, and normal loading. When the helicopter is operated with excessive loads in high density altitude or strong gusty wind conditions, best performance is achieved from a slightly increased airspeed in the descent. For autorotations in light winds, low density altitude, and light loading, best performance is achieved from a slight decrease in normal airspeed. Following this general procedure of fitting airspeed to existing conditions, a pilot can achieve approximately the same glide angle in any set of circumstances and estimate his touchdown point.

When making turns during an autorotative descent, generally use cyclic control only. Use of antitorque pedals to assist or speed the turn causes loss of airspeed and downward pitching of the nose—especially when left pedal is used. When the autorotation is initiated, sufficient right pedal pressure should be used to maintain straight flight and prevent yawing to the left. This pressure should not be changed to assist the turn.

If rotor RPM becomes too high during an autorotative approach, collective pitch should be raised sufficiently to decrease RPM to the normal operating range, then lowered all the way again. This procedure may be repeated as necessary.

> If the throttle has not been closed in practice autorotations, it will be necessary to reduce it as collective pitch is raised; otherwise, the engine and rotor will re-engage. After the collective is lowered, the throttle should be readjusted to maintain a safe engine RPM.

RPM is most likely to increase above the maximum limit during a turn due to the increased back cyclic stick pressure which induces a greater airflow through the rotor system. The tighter the turn and the heavier the gross weight, the higher the RPM will be.

HOVERING AUTOROTATION

As the name implies, hovering autorotations are performed from a hover. They are practiced so that a pilot may become proficient in the proper control technique so that he will automatically make the correct response when confronted with engine stoppage or certain other in-flight emergencies while hovering.

TECHNIQUE:

1. To practice hovering autorotations, establish a normal hovering altitude for the particular helicopter, considering its load and atmospheric conditions, and keep it headed into the wind. Hold maximum allowable RPM.

2. To enter autorotation, close the throttle quickly to ensure a clean split of the needles. This disengages the driving force of the engine from the rotor, thus eliminating torque effect. As throttle is closed, right pedal pressure must be applied to maintain heading. Usually, a slight amount of right cyclic stick will be necessary to keep the helicopter from drifting, but use cyclic control as required to ensure a vertical descent and a level attitude. Leave the collective pitch where it is on entry.

3. In helicopters with low inertia rotor systems, the aircraft will begin to settle immediately. Keep a level attitude and ensure a vertical descent with cyclic control, heading with pedals, and apply upward collective pitch as necessary (generally the full amount is required) to slow the descent and cushion the landing. As upward collective pitch is applied, the throttle will have to be held in the closed position to prevent the rotor from re-engaging.

4. In helicopters with high inertia rotor systems, the aircraft will maintain altitude momentarily after the throttle is closed. Then, as the rotor RPM decreases, the helicopter will start to settle. As it settles, apply upward

collective pitch while holding the throttle in the closed position to slow the descent and cushion the landing. The timing of this collective pitch application and the rate at which it should be applied will depend upon the particular helicopter, its gross weight, and the atmospheric conditions. Cyclic control is used to maintain a level attitude and ensure a vertical descent. Heading is maintained with pedals.

5. When the weight of the helicopter is entirely on the skids, the application of upward collective pitch should be stopped. When the helicopter has come to a complete stop, lower the collective pitch all the way.

6. The timing of the collective pitch is a most important consideration. If it is applied too soon, the remaining RPM will not be sufficient to effect a smooth landing. On the other hand, if collective pitch is initiated too late, ground contact will be made before sufficient blade pitch is available to cushion the landing.

7. When entering the autorotation, the throttle should be rotated to the closed or override position to prevent the engine from re-engaging during the collective pitch application.

COMMON ERRORS:

1. Failing to use sufficient right pedal when power is reduced.

2. Failing to stop all sideward or backward movement prior to touchdown.

3. Failing to time up-collective pitch properly resulting in a hard touchdown.

4. Failing to touch down in a level attitude.

NO-FLARE AUTOROTATION

A no-flare autorotation can be used when the selected landing area is sufficiently long and smooth to permit a ground run. Practice no-flare autorotations may be executed as follows.

TECHNIQUE:

1. When the desired position for initiating the autorotation has been reached, smoothly place the collective pitch stick in the full down position, maintaining cruising RPM with throttle. Decrease throttle to ensure a clean split of the needles and apply sufficient right pedal to maintain the desired heading. After splitting the needles, readjust the throttle so as to keep engine RPM well above normal idling speed but not high enough to cause rejoining of the needles. (The manufacturer will often recommend the RPM to use.)

2. Adjust attitude with cyclic control to obtain the best gliding speed (slowest rate of descent). Be sure to hold

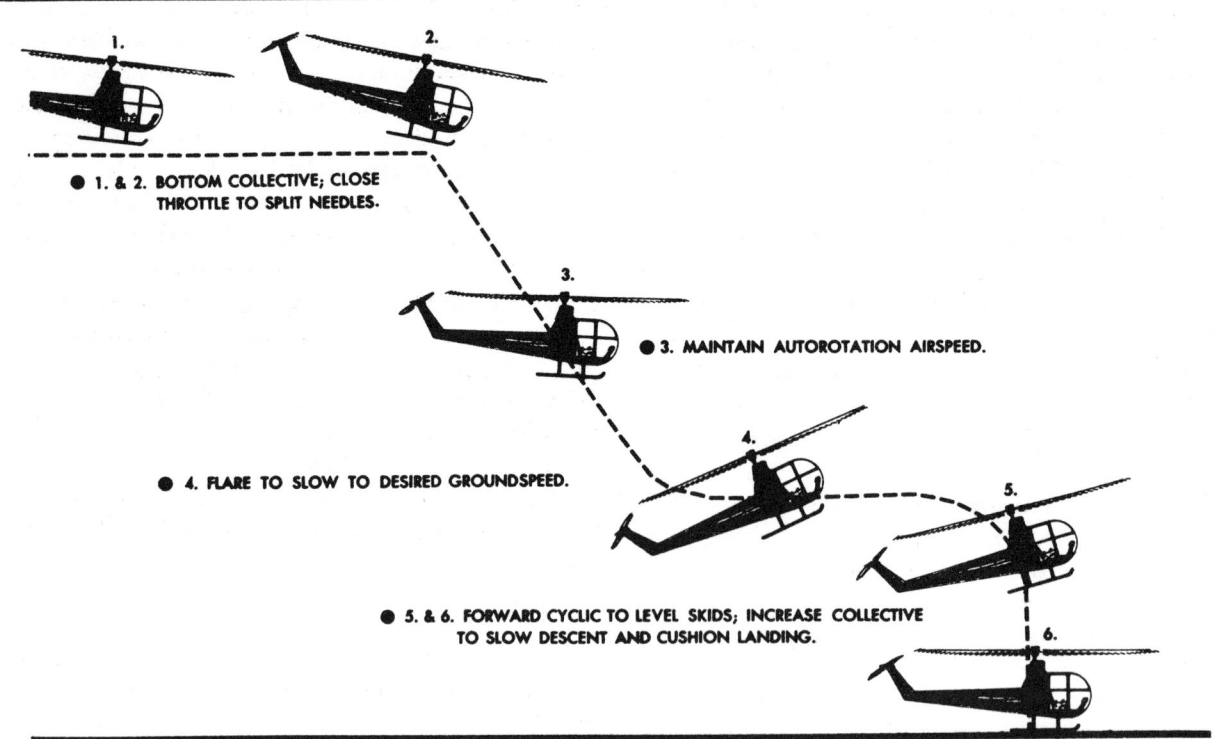

● 1. & 2. BOTTOM COLLECTIVE; CLOSE THROTTLE TO SPLIT NEEDLES.

● 3. MAINTAIN AUTOROTATION AIRSPEED.

● 4. FLARE TO SLOW TO DESIRED GROUNDSPEED.

● 5. & 6. FORWARD CYCLIC TO LEVEL SKIDS; INCREASE COLLECTIVE TO SLOW DESCENT AND CUSHION LANDING.

Figure 77.—Flare autorotation and touchdown.

collective pitch in the full down position. If it is permitted to rise, rotor RPM will decrease due to the increased drag from the increased pitch angle of the blades. At approximately 50 feet above the ground (check the manufacturer's recommendation for each helicopter), raise the nose slightly to obtain the desired landing speed and to slow the rate of descent.

3. If a landing is to be made from the autorotative approach, the throttle should be rotated to the closed or override position at this time and held in this position as collective pitch is raised so that the rotor will not re-engage. As the helicopter approaches normal hovering altitude, maintain a level attitude with cyclic control, maintain heading with pedals, apply sufficient collective pitch (while holding the throttle in the closed position) to cushion the touchdown, and be sure the helicopter is landing parallel to its direction of motion upon contact with the ground. Avoid landing on the heels of the skid gear. The timing of the collective pitch application and the amount applied will be dependent on the rate of descent.

4. After ground contact is made, collective pitch may be increased smoothly (while holding the throttle in the closed position) to keep the helicopter light on the skids and allow it to slow down gradually; or it may be held stationary resulting in a shorter ground run; or it may be lowered cautiously for additional braking action if required due to a fast touchdown and limited landing area. Cyclic control stick should be held slightly forward of neutral and used to maintain direction if landing is made in a crosswind. Pedals should be used to maintain heading. In the event of insufficient pedal to maintain heading control as the rotor RPM decreases after touchdown, cyclic control should be applied in the direction of the turn.

5. After the helicopter has stopped, lower the collective pitch to the full down position.

6. If a power recovery is to be made from practice autorotative approaches, the procedures in (3), (4), and (5) should be replaced with the procedures given under "Power Recovery From Practice Autorotations."

COMMON ERRORS:

1. Failing to use sufficient right pedal when power is reduced.

2. Lowering the nose too abruptly when power is reduced, thus placing the helicopter in a dive.

3. Failing to maintain full down collective pitch during the descent.

4. Application of up-collective pitch at an excessive altitude resulting in a hard landing, loss of heading control, and possible damage to the tail rotor and to the main rotor blade stops.

5. Pulling the nose up just prior to touchdown.

FLARE AUTOROTATION

A flare autorotation (fig. 77) enables the pilot to land a helicopter at any speed between that resulting in little or no landing run up to that of a no-flare autorotation; that is, anywhere between a zero groundspeed and the speed of touchdown resulting from a no-flare autorotation. The speed at touchdown and the resulting ground run will depend on the rate and amount of the flare—the greater the degree of flare and the longer it is held, the slower the touchdown speed and the shorter the ground run. The slower the speed desired at touchdown from an autorotation, the more accurate must be the timing and speed of the flare, especially in helicopters with low inertia rotor systems.

TECHNIQUE:

1. Enter the flare autorotation in the same manner as the no-flare autorotation. The technique is the same down to the point where the flare is to begin. This point is slightly lower than the point at which the nose is raised in the no-flare autorotation.

2. At appoximately 35 to 60 feet above the ground, depending on the helicopter (check the manufacturer's recommendations), initiate the flare by moving the cyclic stick smoothly to the rear. Heading is maintained by the pedals. Care must be exercised in the execution of the flare so that the cyclic control is not moved rearward so abruptly as to cause the helicopter to climb, nor should it be moved so slowly as to allow the helicopter to settle so rapidly that the tail rotor might strike the ground. As forward motion decreases to the desired groundspeed, move cyclic control forward to level the helicopter in preparation for landing. (If a landing is to be made, the throttle should be rotated to the closed or override position at this time; if a power recovery is to be made, it should be made as the helicopter reaches the level attitude.)

3. The altitude at this time should be approximately 3 to 10 feet, depending on the helicopter. If a landing is to be made, allow the helicopter to descend vertically. Apply collective pitch smoothly as necessary to check the descent and cushion the landing. As collective pitch is increased, hold the throttle in the closed position so that the rotor will not re-engage. Additional right pedal is required to maintain heading as collective pitch is raised due to the reduction in rotor RPM and the resulting reduced effect of the tail rotor.

4. After touchdown and the helicopter has come to a complete stop, lower the collective pitch to the full down position.

COMMON ERRORS:

1. Failing to use sufficient right pedal when power is reduced.

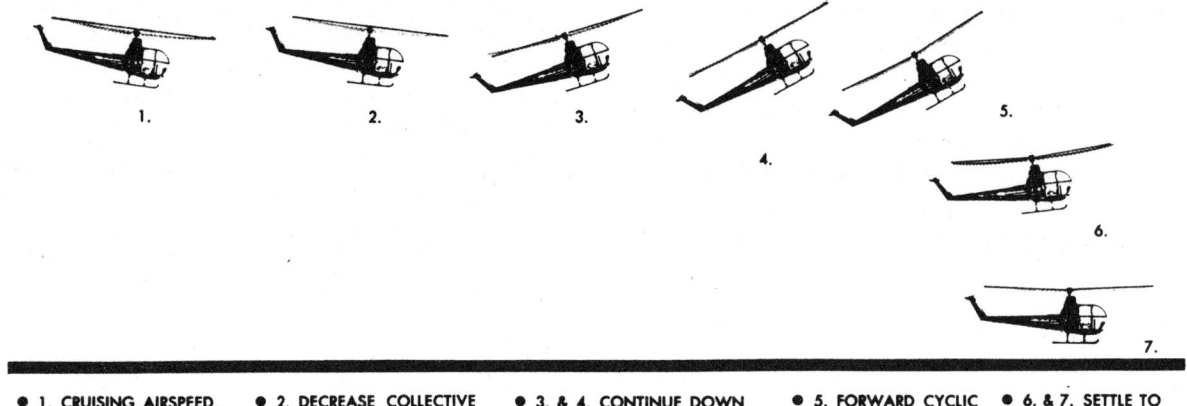

● 1. CRUISING AIRSPEED, SAFE ALTITUDE.   ● 2. DECREASE COLLECTIVE SLIGHT BACK CYCLIC.   ● 3. & 4. CONTINUE DOWN COLLECTIVE AND BACK CYCLIC TO DECREASE GROUNDSPEED.   ● 5. FORWARD CYCLIC UP COLLECTIVE.   ● 6. & 7. SETTLE TO A NORMAL HOVER.

FIGURE 78.—Rapid deceleration or quick stop.

2. Lowering the nose too abruptly when power is reduced, thus placing the helicopter in a dive.

3. Failing to maintain full down-collective pitch during the descent.

4. Application of up-collective pitch at an excessive altitude resulting in a hard landing, loss of heading control, and possible damage to the tail rotor and to the main rotor blade stops.

5. Applying up-collective pitch before a level attitude is obtained. (If timing is late, it may be necessary to apply up-collective before a level attitude is obtained.)

6. Pulling the nose up just prior to touchdown on full autorotation.

POWER RECOVERY FROM PRACTICE AUTOROTATIONS

A power recovery is used to terminate practice autorotations at a point prior to actual touchdown. After the power recovery, a landing can be made or a go-around initiated.

TECHNIQUE:

1. To effect a power recovery after the flare or level-off from an autorotation, coordinate upward collective pitch and increased throttle to join the needles at operating RPM. The throttle and collective pitch must be coordinated properly. If the throttle is increased too fast or too much, an engine overspeed will occur; if throttle is increased too slowly or too little in proportion to the increase in collective pitch, a loss of rotor RPM will result. Use sufficient

collective pitch to check the descent and coordinate left pedal pressure with the increase in collective pitch to maintain heading.

2. If a go-around is to be made, the cyclic control should be moved forward smoothly to re-enter forward flight. If a landing is to be made following the power recovery, the helicopter can be brought to a hover at normal hovering altitude.

3. In transitioning from a practice autorotation to a go-around, care must be exercised to avoid an altitude–airspeed combination which would place the helicopter in an unsafe area of the height–velocity chart for that particular helicopter.

COMMON ERRORS:

1. Initiating recovery too late requiring a rapid application of controls and resulting in overcontrolling.

2. Failing to obtain and maintain a level attitude near the ground.

3. Adding throttle before the collective pitch.

4. Failing to coordinate throttle and collective pitch properly resulting in an engine overspeed or loss of RPM.

5. Failing to coordinate left pedal with the increase in power.

RAPID DECELERATION OR QUICK STOP

Although used primarily for coordination practice, decelerations (fig. 78) can be used to effect a quick stop in the air. The purpose of the maneuver is to maintain a constant altitude, heading, and RPM while slowing the helicopter to a desired groundspeed. The maneuver re-

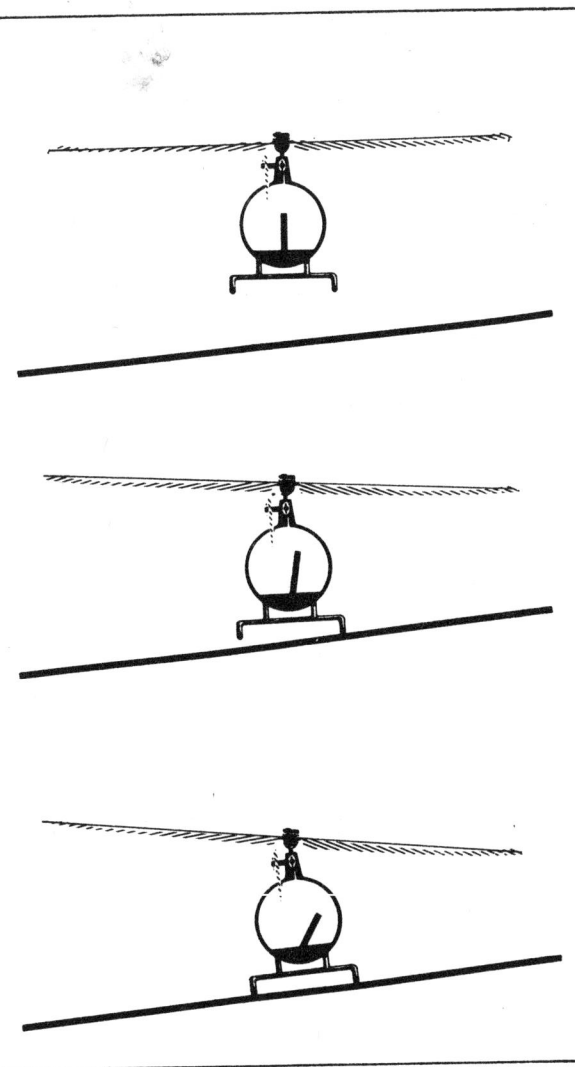

FIGURE 79.—Slope landing or takeoff.

quires a high degree of coordination of all controls. It is practiced at an altitude that will permit a safe clearance between tail rotor and ground throughout the maneuver, especially at the point where the pitch attitude is highest. The altitude at completion should be no higher than the maximum safe hovering altitude prescribed by the manufacturer. In selecting an altitude at which to begin the maneuver, the overall length of the helicopter and the height–velocity chart must be considered.

Although the maneuver is called a rapid deceleration or quick stop, this does not mean that it should be rushed to completion. The rate of deceleration is at the discretion of the pilot. A quick stop is completed when the helicopter comes to a hover during the recovery.

TECHNIQUE:

1. Begin the maneuver at a fast hover speed headed into the wind at an altitude high enough to avoid danger to the tail rotor during the flare but low enough to stay out of the height-velocity chart shaded area throughout the performance. This altitude should be low enough that the helicopter can be brought to a hover during the recovery.

2. The deceleration is initiated by applying aft cyclic to reduce forward speed. Simultaneously, the collective pitch should be lowered as necessary to counteract any climbing tendency. The timing must be exact. If too little down collective is applied for the amount of aft cyclic, a climb will result. Conversely, if too much down collective is applied for the amount of aft cyclic, a descent will result. The more rapid the application of aft cyclic, the more rapid the down collective application. As collective pitch is lowered, right pedal should be increased to maintain heading and throttle should be adjusted to maintain RPM.

3. After speed has been reduced to the desired amount, recovery is initiated by lowering the nose and allowing the helicopter to descend to a normal hovering altitude in level flight and zero groundspeed. During the recovery, collective pitch should be increased as necessary to stop the helicopter at normal hovering altitude; throttle should be adjusted to maintain RPM; and left pedal pressure should be increased as necessary to maintain heading.

COMMON ERRORS:

1. Initiating the maneuver with down collective.

2. Applying aft cyclic stick too rapidly initially, causing the helicopter to "balloon" (a sudden gain in altitude).

3. Failing to effectively control the rate of deceleration to accomplish the desired results.

4. Allowing the helicopter to stop forward motion in a tail-low attitude.

5. Failing to maintain proper RPM.

SLOPE OPERATIONS

The approach to a slope is not materially different from the approach to any other landing area. During slope operations, allowance must be made for wind, barriers, and forced landing sites in case of engine failure. Since the slope may constitute an obstruction to wind passage, turbulence and downdrafts must be anticipated.

Slope landing

It is usually best to land the helicopter cross-slope rather than upslope. Landing downslope or downhill is not recommended because of the possibility of striking the tail rotor on the ground.

TECHNIQUE:

1. At the termination of the approach, move slowly toward the slope being especially careful not to turn the tail upslope. The helicopter should be hovered in position cross-slope over the spot of intended landing (fig. 79).

2. A slight downward pressure on the collective pitch will start the helicopter descending slowly. As the up-slope skid touches the ground, apply cyclic stick pressure in the direction of the slope. This will hold the skid against the slope while the helicopter is continuing to be let down with the collective pitch.

3. As collective pitch is reduced, continue to move the cyclic stick toward the slope to maintain a fixed position. The slope must be shallow enough to allow the pilot to hold the helicopter against it with the cyclic stick during the entire landing. (A slope of 5° is considered maximum for normal operation of most helicopters. Each make of helicopter will generally have its own peculiar way of indicating to the pilot when he is about to run out of lateral cyclic stick such as the rotor hub hitting the rotor mast, vibrations felt through the cyclic stick, and others. A landing should not be made in these instances since this indicates to the pilot that the slope is too steep.)

4. After the downslope skid is on the ground, continue to lower the collective pitch all the way to the bottom. Normal operating RPM should be maintained until the full weight of the helicopter is on the skids. This will assure adequate RPM for immediate takeoff in case the helicopter should start to slide down the slope. Pedals should be used as necessary throughout the landing to maintain heading.

COMMON ERRORS:

1. Failure to maintain proper RPM throughout the entire maneuver.

2. **Letting the helicopter down too rapidly.**

Slope takeoff

The procedure for a slope takeoff is almost the exact reverse of that for a slope landing (fig. 79).

TECHNIQUE:

1. Adjust throttle to obtain takeoff RPM and move the cyclic stick in the direction of the slope so that the rotor rotation is parallel to the true horizontal rather than the slope.

2. Apply up-collective pitch smoothly. As the helicopter becomes light on the skids, apply pedal as needed to maintain heading.

3. When the downslope skid has risen and the helicopter approaches a level attitude, move the cyclic stick back to the neutral position. Continue to apply up-collective pitch and take the helicopter straight up to a hover before moving away from the slope. In moving away from the slope, the tail should not be turned upslope because of the danger of the tail rotor striking the ground.

COMMON ERRORS:

1. **Failure to adjust cyclic stick to keep the helicopter from sliding downslope.**

2. Failure to maintain proper RPM.

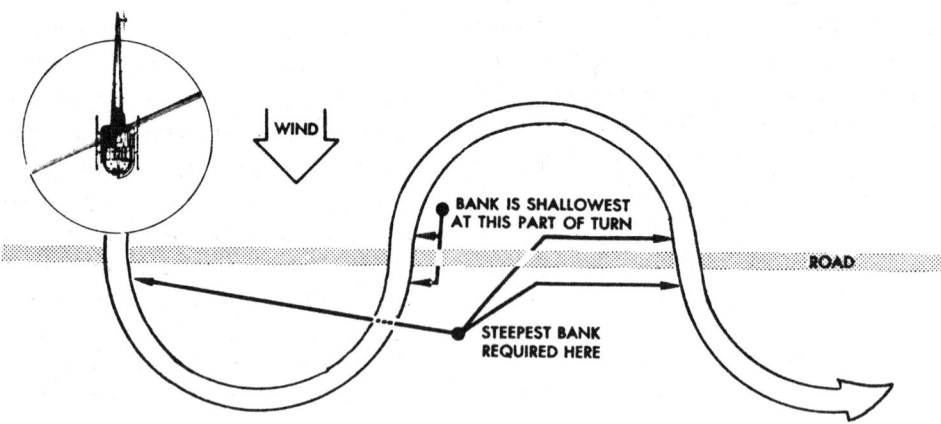

FIGURE 80.—"S" turns.

RECOVERY FROM LOW ROTOR RPM

Recovery from low rotor RPM is a procedure used to return to the normal rotor operating RPM. This recovery procedure, if performed properly, will normally regain lost rotor RPM while still maintaining flight. A low rotor RPM condition is the result of having an angle of attack on the main rotor blades (induced by too much upward collective pitch) that has created a drag so great that engine power available, or being utilized, is not sufficient to maintain normal rotor operating RPM.

When a low rotor RPM condition is realized, immediately lower the collective pitch. This action will decrease the angle of attack of the main rotor blades which, because of reduced rotor drag, will momentarily relieve excessive engine load. This action will also cause the helicopter to settle because some lift will be lost. As the helicopter begins to settle, smoothly raise the collective pitch just enough to stop the settling motion. This procedure, under critical conditions, may have to be repeated several times to regain normal rotor RPM. When operating at sufficient altitudes above the surface, however, it may be necessary to lower the collective pitch only once to regain sufficient rotor RPM. The *amount* that the collective pitch can be lowered will depend on the altitude available at the time the low rotor RPM condition occurs. When hovering near the surface, the collective pitch should be lowered cautiously to preclude hard contact with the terrain. When the rotor RPM begins to rise and attains approximately normal rotor operating RPM, anticipate decreasing the throttle slowly to prevent the engine from overspeeding.

If recovery from a low rotor RPM condition is not effected soon enough, lifting power of the main rotor blades will be lost, as well as pedal control. Pedal control loss occurs as a result of the loss of tail rotor RPM because the tail rotor is driven by the main rotor and its RPM is directly proportional to the main rotor RPM. If pedal control is lost and the altitude is such that a landing can be effected before the turning rate increases dangerously, decrease collective pitch slowly, maintain a level attitude with cyclic control, and land.

"S" TURNS

"S" turns present one of the most elementary problems in the practical application of the turn and in the correction for wind drift in turns. The reference line used, whether a road, railroad, or fence, should be straight for a considerable distance and should extend as nearly perpendicular to the wind as possible.

The object of "S" turns is to fly a pattern of two half circles of equal size on opposite sides of the reference line

(fig. 80). The maneuver should be started at an altitude of 500 feet above the terrain and a constant altitude maintained throughout the maneuver. "S" turns may be started at any point. However, during early training, it may be beneficial to start on a downwind heading. Entering downwind permits the immediate selection of the steepest bank that is desired throughout the maneuver. The discussion that follows is based on choosing a reference line that is perpendicular to the wind and starting the maneuver on a downwind heading.

As the helicopter crosses the reference line, a bank is immediately established. This initial bank will be the steepest used throughout the maneuver since the helicopter is headed directly downwind. The bank should be gradually reduced as necessary to describe a ground track in a half circle. The turn should be timed so that, as the rollout is completed, the helicopter is crossing the reference line perpendicular to it and headed directly upwind. A bank is immediately entered in the opposite direction to begin the second half of the "S". Since the helicopter is on an upwind heading, this bank (and the one just completed before crossing the reference line) will be the shallowest in the maneuver. It should be gradually increased as necessary to describe a ground track which is a half circle identical in size to the one previously completed on the other side of the reference line. The steepest bank in this turn should be attained just prior to rollout (when the helicopter is approaching the reference line nearest to a downwind heading). This bank, along with the initial bank entered at the beginning of the maneuver, will be the steepest bank used in "S" turns. The turn should be timed so that, as the rollout is completed, the helicopter is crossing the reference line perpendicular to it and again headed directly downwind.

As a summary, the angle of bank required at any given point in the maneuver is dependent on the groundspeed—the faster the groundspeed, the steeper the bank; the slower the groundspeed, the shallower the bank. Or, to express it another way, the more nearly the helicopter is to a downwind heading, the steeper the bank; the more nearly it is to an upwind heading, the shallower the bank.

In addition to varying the angle of bank to correct for drift in order to maintain the proper radius of turn, the helicopter must also be flown with a drift correction angle (crab) in relation to its ground track, except, of course, when it is on direct upwind or downwind headings or there is no wind. One would normally think of the fore and aft axis of the helicopter as being tangent to the ground track pattern at each point. However, this is not the case. During the turn on the upwind side of the reference line (side from which the wind is blowing), the nose of the helicopter will be crabbed toward the outside of the circle. During the turn on the downwind side of the reference line (side of the reference line opposite to the direction from which the wind is blowing), the nose of the helicopter will be crabbed toward the inside of the circle. In either case, it is obvious that the helicopter is being crabbed into the wind just as it is when trying to maintain a straight ground track. The amount of crab depends upon the wind velocity and how nearly the helicopter is to a crosswind position. The stronger the wind, the greater the crab angle at any given position for a turn of a given radius. The more nearly the helicopter is to a crosswind position, the greater the crab angle. The maximum crab angle should be at the point of each half circle farthest from the reference line.

A standard radius for S turns cannot be specified. This radius will depend on the airspeed of the helicopter, the velocity of the wind, and the initial bank chosen for entry.

Chapter 12. CONFINED AREA, PINNACLE, AND RIDGELINE OPERATIONS

A confined area is an area where the flight of the helicopter is limited in some direction by terrain or the presence of obstructions, natural or manmade. For example, a clearing in the woods, a city street, a road, a building roof, and so on, can each be regarded as a confined area.

Barriers on the ground and the ground itself may interfere with the smooth flow of air, resulting in turbulence. This interference is transmitted to upper air levels as larger but less intense disturbances. Therefore, the greatest turbulence is usually found at low altitudes. Gusts are unpredictable variations in wind velocity. Ordinary gusts are dangerous only in slow flight at very low altitudes. The pilot may be unaware of the gust and its cessation may reduce airspeed below that required to sustain flight due to the loss in effective translational lift. Gusts cannot be planned for or anticipated. Turbulence, however, can generally be predicted. Turbulence will be found in the following areas when wind velocity exceeds 10 MPH (fig. 81):

1—Near the ground on the downwind side of trees, buidings, hills, or other obstructions. The turbulent area is always relative in size to that of the obstacle, and relative in intensity to the velocity of the wind.

2—Near the ground on the immediate upwind side of any solid barrier such as trees in leaf and buildings. This condition is not generally dangerous unless the wind velocity is approximately 20 MPH or higher.

3—In the air, above and slightly downwind of any sizable obstruction, such as a hill or mountain range. The size of the obstruction and the wind velocity govern the height to which the turbulence extends and also its severity.

GENERAL RULES FOR CONFINED AREA OPERATIONS

Some general rules can be stated that apply to helicopter operations in any type of confined area. The following are some of the more important ones to consider regardless of whether such areas are enclosed, or are slopes or pinnacles.

1. Know the direction and approximate speed of the wind at all times and plan landings and takeoffs with these wind conditions in mind. This does not necessarily mean that takeoffs and landings will always be made into the wind, but wind must be considered, and many times its velocity will determine proper avenues of approach and takeoff.

2. If possible, plan a flight path over areas suitable for forced landings in case of engine failure. It may be necessary to choose between an approach which is crosswind but over an open area and one directly into the wind but over heavily wooded or extremely rough terrain where a safe forced landing would be impossible. Perhaps the initial phase of the approach can be made crosswind over the open area and then it may be possible to execute a turn into the wind for the final portion of the approach.

3. Always operate the helicopter as closely to its normal capabilities as possible considering the situation at hand. In all confined area operations, with the exception of the pinnacle operation, the angle of descent should be no steeper than is necessary to clear any barrier in the avenue of approach and still land on the selected spot. The angle of climb, on takeoff, should not be steeper than is necessary to clear any barrier. It is better to clear the barrier by a few feet and maintain normal operating RPM, with perhaps a reserve of power, than it is to clear the obstruction by a wide margin but with a dangerously low RPM and no power reserve.

4. Always make the landing to a specific point and not to just some general area. The more confined the area, the more essential it is that the helicopter be landed precisely at a definite point. This spot must be kept in sight during the entire final approach.

5. Any material increase in elevation between the point of takeoff and the point of intended landing must be given due consideration as sufficient power must be available to bring the helicopter to a hover at the point of the intended landing. A decrease in wind should also be allowed for because of the presence of obstructions.

6. When flying a helicopter near obstructions, always consider the tail rotor. A safe angle of descent over barriers must be established to ensure tail rotor clearance of

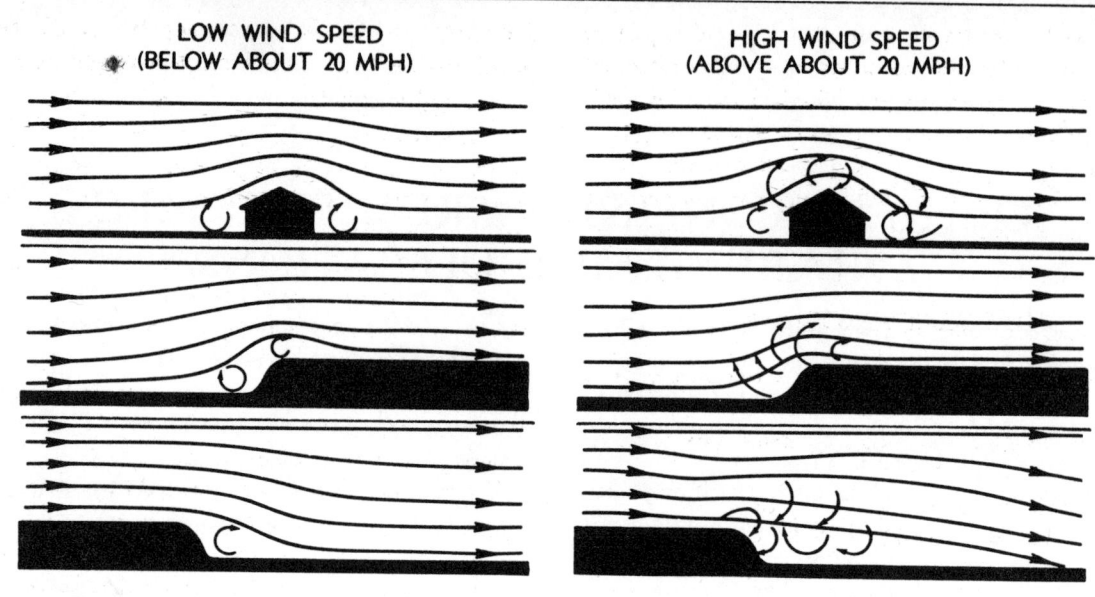

FIGURE 81.—Eddy currents are formed when the wind flows over uneven ground or obstructions.

all obstructions. After coming to a hover, care must be taken to avoid swinging the tail into obstructions.

7. If possible, a normal takeoff from a hover should be made when departing a confined area. However, if barriers of sufficient height exist that would make this impossible, then a maximum performance takeoff should be made.

PINNACLE AND RIDGELINE OPERATIONS

A pinnacle is an area from which the ground drops away steeply on all sides. A ridgeline is a long area from which the ground drops away steeply on one or two sides, such as a bluff or precipice. Barriers are not usually present on pinnacles or ridgelines; but if they are, a combination of pinnacle and confined area operations may be necessary when operating into and out of such areas. That is, an area may require a pinnacle-type operation during the approach and landing, but if the strength of the wind dictates the takeoff path and barriers exist under that path, a confined area-type takeoff may be required when departing that area. Conversely, conditions and terrain may justify a confined area-type approach into an area and a pinnacle-type departure from that area.

The absence of barriers does not necessarily lessen the difficulty of pinnacle or ridgeline operations. Updrafts, downdrafts, and turbulence, together with unsuitable terrain in which to make a forced landing, may still present extreme hazards.

General rules for pinnacle and ridgeline operations

The following are some of the more important rules to consider when conducting pinnacle or ridgeline operations:

1. If necessary to climb to a pinnacle or ridgeline, the climb should be performed on the upwind side, when practicable, to take advantage of any updrafts.

2. Load, altitude, wind conditions, and terrain features determine the angle to use in the final part of an approach. A steeper-than-normal approach may be used when barriers or excessive downdrafts exist. A shallower-than-normal approach may be used when there are no barriers or downdrafts and when it is suspected that the helicopter cannot be hovered out of ground effect. In this case, an approach to the ground may be necessary.

3. The approach path to a ridgeline is usually parallel to that ridgeline and as nearly into the wind as possible. If a crosswind exists, remain clear of downdrafts on the leeward or downwind side of the ridgeline. If the wind velocity makes the crosswind landing hazardous, a low coordinated turn into the wind may be made just prior to terminating the approach.

4. When making an approach to a pinnacle, avoid leeward turbulence and keep the helicopter within reach of a forced landing area as long as possible.

5. Since pinnacles and ridgelines are generally higher than the immediate surrounding terrain, gaining airspeed on the takeoff is more important than gaining altitude.

The airspeed gained will cause a more rapid departure from the area. In addition to covering unfavorable terrain rapidly, a higher airspeed affords a more favorable glide angle and thus contributes to the chances of reaching a safe area in the event of a forced landing. If no suitable area is available, a higher airspeed will permit a more effective flare prior to making an autorotative landing.

HIGH RECONNAISSANCE

The primary purpose of the high reconnaissance is to determine the suitability of an area for a landing. In a high reconnaissance, the following items should be accomplished:

1—Determine wind direction and speed.

2—Select the most suitable flight paths in and out of the area, with particular consideration being given forced landing areas.

3—Plan the approach and select a point for touchdown.

4—Locate and determine the size of barriers, if any, immediately around the area.

A high reconnaissance is flown at approximately 500 feet; however, a higher altitude may be required in some helicopters. A general rule to follow is to ensure that sufficient altitude is available at all times to land into the wind in case of engine failure. This means the greatest altitude will be required when headed downwind. A complete circle of the area should be made if possible. A 45° angle of observation will generally allow the best estimate of the height of barriers, the presence of obstacles, the size of the area, and the slope of the terrain. Safe altitudes and airspeeds should be maintained and a forced landing area should be kept within reach whenever possible.

The approach path should be generally into the wind and over terrain that minimizes the time that the helicopter is not in reach of a forced landing area. If by flying at an angle to the wind, a forced landing area can be kept in reach, then do so. It is more important to have an available forced landing area than to fly directly into the wind, especially if the wind is not too strong. The decision should be made as to the type of approach. If at all possible, a normal approach should be made; however, if there are high barriers, a steeper approach will be required.

LOW RECONNAISSANCE

In the low reconnaissance, verify what was seen in the high reconnaissance and pick up anything new that may have been missed. Check especially for wires, slopes, and small crevices because these are especially difficult to see from a higher altitude.

A low reconnaissance begins shortly after entry to the approach and ends at touchdown. During this time, objects on the ground can be better identified and the height of barriers, if any, better estimated. The view of the approach path is greatly improved. The approach should be as close to normal as possible. If new information warrants a change in flight path or angle of descent, it should be made; however, if a major change in angle of descent is required, a go-around should be made. If a decision to go around is made, it should be done prior to losing effective translational lift.

If a decision is made to complete the approach, the termination should normally be to a hover so the landing spot can be carefully checked before the landing is made. Under certain conditions, however, it may be desirable to terminate the approach to the ground. Whether terminating at a hover or on the ground, once the helicopter is on the surface, operating RPM should be maintained until the stability of the helicopter is checked to be sure of a secure and safe position.

GROUND RECONNAISSANCE

Before takeoff, a ground reconnaissance is made to determine the type takeoff to be performed, to determine the point from which the takeoff should be initiated to ensure the maximum amount of available area and, finally, how to best get the helicopter from the landing spot to the proposed takeoff position.

The first thing to check is the wind. If the rotor is left turning, walk a sufficient distance from the helicopter to ensure that the downwash of the blades does not interfere. Light dust or grass may be dropped and the direction observed in which they are blown.

The next step is to go to the downwind end of the available area and mark a position for takeoff so that the tail and main rotors will have sufficient clearance from the obstructions, if any, behind the helicopter. A sturdy marker such as a heavy stone or log should be used so that it will not blow away.

If rearward flight is required to reach the takeoff position, reference markers should be placed in front of the helicopter in such a way that a ground track can be safely followed to the takeoff position and so the pilot can see the marker for the takeoff position without going beyond it. If wind conditions and available area permit, however, the helicopter should be slowly hovered downwind from the landing position to the takeoff position.

REFERENCES FOR ADDITIONAL STUDY

The following publications are available from federal government sources and are recommended for additional study by applicants for a helicopter class rating.
1. Federal Aviation Regulations
2. Airman's Information Manual (AIM)
3. National Transportation Safety Board Procedural Regulations, Part 430. May be ordered, without charge, from: National Transportation Safety Board, Publications Branch, 800 Independence Ave., SW., Washington, D.C. 20591
4. Aviation Weather, AC 00-6
5. Pilot's Handbook of Aeronautical Knowledge, AC 61-23A
6. VFR Exam-O-Grams

You may visit any General Aviation District Office for further suggestions pertaining to additional study material, or you may refer to Federal Aviation Administration *Advisory Circular Checklist and Status of Regulations,* AC 00-2 (latest revision), for a listing of FAA publications. This circular is free and can be obtained by writing to:

Department of Transportation
Distribution Unit, TAD 484.3
Washington, D.C. 20590

FAA-approved helicopter flight manuals can be obtained from individual aircraft manufacturing companies or from local dealers and distributors. VFR Exam-O-Grams can be obtained, without charge, from the FAA Aeronautical Center, Flight Standards Technical Division, Operations Branch, P.O. Box 25082, Oklahoma City, Oklahoma 73125.

GLOSSARY

advancing blade—That half of the rotor disc in which the rotation of the blade is moving in the same direction as the movement of the helicopter. If the helicopter is moving forward, the advancing blade will be in the right half of the rotor disc; if moving backward, it will be in the left half; if moving sideward to the left, it will be in the forward half; and if moving sideward to the right, it will be in the rear half.

airfoil—Any surface designed to obtain a useful reaction from the air through which it moves in the form of lift.

angle of attack—The acute angle measured between the chord of an airfoil and the relative wind.

articulated rotor—A rotor system in which the blades are free to flap, drag, and feather.

blade damper—A device (spring, friction, or hydraulic) installed on the vertical (drag) hinge to diminish or dampen blade oscillation (hunting) around this hinge.

blade loading—The load placed on the rotor blades of a helicopter, determined by dividing the gross weight of the helicopter by the combined area of all the rotor blades.

center of gravity—An imaginary point where the resultant of all weight forces in the body may be considered to be concentrated for any position of the body.

center of pressure—The imaginary point on the chord line where the resultant of all aerodynamic forces of an airfoil section may be considered to be concentrated.

centrifugal force—The force created by the tendency of a body to follow a straight-line path against the force which causes it to move in a curve, resulting in a force which tends to pull away from the axis of rotation.

chord—An imaginary straight line between the leading and trailing edges of an airfoil.

collective pitch control—The method of control by which the pitch of all rotor blades is varied equally and simultaneously.

coriolis effect—The tendency of a mass to increase or decrease its angular velocity when its radius of rotation is shortened or lengthened, respectively.

cyclic pitch control—The control which changes the pitch of the rotor blades individually during a cycle of revolution to control the tilt of the rotor disc, and therefore, the direction and velocity of horizontal flight.

delta hinge (flapping hinge)—The hinge with its axis parallel to the rotor plane of rotation, which permits the rotor blades to flap to equalize lift between the advancing blade half and retreating blade half of the rotor disc.

density altitude—Pressure altitude corrected for temperature and humidity.

disc area—The area swept by the blades of the rotor. This is a circle with its center at the hub axis and a radius of one blade length.

disc loading—The ratio of helicopter gross weight to rotor disc area (total helicopter weight divided by the rotor disc area).

dissymmetry of lift—The unequal lift across the rotor disc resulting from the difference in the velocity of air over the advancing blade half and retreating blade half of the rotor disc area.

feathering axis—The axis about which the pitch angle of a rotor blade is varied. Sometimes referred to as the spanwise axis.

feathering action—That action which changes the pitch angle of the rotor blades periodically by rotating them around their feathering (spanwise) axis.

flapping—The vertical movement of a blade about a delta (flapping) hinge.

freewheeling unit—A component part of the transmission or power train which automatically disconnects the main rotor from the engine when the engine stops or slows below the equivalent of rotor RPM.

ground effect—The cushion of denser air confined beneath the rotor system of a hovering helicopter which gives additional lift and thus decreases the power required to hover.

gyroscopic precession—A characteristic of all rotating bodies. When a force is applied to the periphery of a rotating body parallel to its axis of rotation, the rotating body will tilt in the direction of the applied force $90°$ later in the plane of rotation.

hovering in ground effect—Maintaining a fixed position over a spot on the ground or water which compresses a cushion of high-density air between the main rotor and the ground or water surface and thus increases the lift produced by the main rotor. Normally the main rotor must be within one-half rotor diameter to the ground or water surface in order to produce an efficient ground effect.

hovering out of ground effect—Maintaining a fixed position over a spot on the ground or water at some altitude above the ground at which no additional lift is obtained from ground effect.

hunting—The tendency of a blade (due to coriolis effect) to seek a position ahead of or behind that which would be determined by centrifugal force alone.

pitch angle—The angle between the chord line of the rotor blade and the reference plane of the main rotor hub or the rotor plane of rotation.

rigid rotor—A rotor system with blades fixed to the hub in such a way that they can feather but cannot flap or drag.

semirigid rotor—A rotor system in which the blades are fixed to the hub but are free to flap and feather.

slip—The controlled flight of a helicopter in a direction not in line with its fore and aft axis.

solidity ratio—The ratio of total rotor blade area to total rotor disc area.

standard atmosphere—Atmospheric conditions in which (1) the air is a dry, perfect gas; (2) the temperature at sea level is 59° F. (15° C.); (3) the pressure at sea level (or reduced to sea level) is 29.92 inches of Hg; and (4) the temperature gradient is approximately 3.5° F. per 1,000-foot change in altitude.

tip-path plane—The plane in which rotor blade tips travel when rotating.

tip speed—The rotative speed of the rotor at its blade tips.

tip stall—The stall condition on the retreating blade which occurs at high forward airspeeds.

torque—A force or combination of forces that tends to produce a countering rotating motion. In a single rotor helicopter where the rotor turns counterclockwise, the fuselage tends to rotate clockwise (looking down on the helicopter).

translational lift—The additional lift obtained through airspeed because of increased efficiency of the rotor system whether it be when transitioning from a hover into horizontal flight or when hovering in a wind.

INDEX

U. S. GOVERNMENT PRINTING OFFICE : 1977 O - 238-573

41 L